The Singularity Shift

Other books written by Dave Karpinsky

Artificial Intelligence & Information Technology
- Artificial Intelligence (AI) for Daily Life: A Practical Guide to Artificial Intelligence
- AI and Creativity: How Machines are Changing Art, Music & Literature
- AI-Powered PM: Leveraging Artificial Intelligence for Enhanced Efficiency and Success
- Artificial Intelligence Rise and Humanity Fall
- Data-Driven Future: Harnessing AI and Big Data for Tomorrow's Challenges
- Deepfake Technology: The Dark Side of AI, Manipulation and Digital Deception.
- Fixing Failed Projects: How to Master the Art of Project Turnaround
- From Data to Decisions: The Role of AI in Business Intelligence
- Jobs AI Will Replace: Re-tool or Be Left Behind
- Mastering Advanced Project Management: Strategies for Excellence
- Mastering Project Management: In complex, stressful & high-pressure environments
- SAP S/4 Implementation: A Comprehensive Guide for Practitioners
- SAP S/4 Implementation Methodologies
- SAP S/4 Implementation – Volume 1: Prep & Explore Phases
- SAP S/4 Implementation – Volume 2: Realize & Deploy Phases
- SAP S/4 Implementation – Volume 3: When Projects Fail
- The Five-Day Organizational Change Manager
- The Five--Day Project Manager
- The Project Management Masterclass: Advanced Techniques for Success

- The Rise of Real-Time Analytics: Speed, Precision, and Competitive Edge

Business & Finance
- Building Wealth in Developing Nations: A Comprehensive Step-by-Step Guide to Empower Emerging Markets
- Chief Executive's (CxO) Playbook: The First 90 Days Guide to Success
- Creating a Deployment Plan: Navigating Complexity to Deliver Success
- Creating a Strategic Roadmap: Crafting the Blueprint from Vision to Execution
- Investing Strategies of the Rich and Famous: Discover How to Diversify Your Portfolio for Maximum Returns
- Outsmart the Game: Winning When the Rules Are Rigged
- The Data Delusion: Exposing False Metrics That Shape Your World
- Trust is the New Currency: How Connection Wins in the Age of AI

Life Coach & Mentor Series
- Aspiring Entrepreneurs
- Bored Housewife
- Career Transition
- Couples and Relationships
- Mid-Life Crisis
- Mindful Healthy Living
- Project Managers
- Seeking Life's Purpose
- Surviving Holidays with In-laws

Science & Physics
- Game Over. Reset Earth
- Quantum Entanglement: The God Effect and the Secrets of Reality
- Multiverse Parallel Dimensions: The Theories and Possibilities of Parallel Universes
- Space-Time Continuum: Navigating the Quantum of the Fourth Dimension
- The Hubble Tension: The Universe's Expansion, Cosmology Crisis, and the Limits of the Big Bang Theory
- The Singularity Shift: Unveiling the Future of Humanity and Intelligence
- Twin Paradox: Solving the Puzzle of Special Relativity

Sociology & Politics
- America at War: Russia, China, Iran, S Korea
- Blue Zones Volume 1: Mystery and Science of Blue Zones
- Blue Zones Volume 2: Longevity Lessons of Blue Zones
- Decline of American Supremacy: Understanding the Erosion, Shaping the Future
- Future of Military Technology Powered by AI: How countries are transforming their warfare
- Herd Instinct: Understanding the Human Psychology of Collective Behavior
- Our Idiot Species: Evolution in Reverse
- Preventing Squatters: A Comprehensive Guide to Protecting Your Property
- Puppet Masters: The Hidden Hands of Political Power
- The Great War of China vs Russia: A Future Battlefield that Reshapes the World

- The Modern Stoic: 365 Ancient Practices for Wisdom, Peace, Purpose ad Strength
- The Next Battlefield: How AI, Robotics, and Biotechnology are Transforming Warfare
- The Savage Guide to Winning: The Brutal Truth About Success
- The Trump Effect: Return to the White House
- The Vatican Murder Cover-Up
- Unf*k Yourself: A No-Bullsh!t Guide to Taking Control
- Warfare Redefined: Military Technologies and Tactics of Tomorrow's Superpowers
- Zero F*cks Given: How to Stop Worrying and Live Your Life
- God & AI Series:
    - o Is There God: According to Artificial Intelligence (AI)
    - o What is God: According to Artificial Intelligence (AI)
    - o What is God's Plan: According to Artificial Intelligence (AI)

"The machines aren't becoming human—the humans are becoming machines, one app at a time.
— *Dave Karpinsky*

The Singularity Shift

Unveiling the Future of

Humanity and Intelligence

Dave Karpinsky, PhD, MBA, PMP, Prosci

Green Parrot Media

Contents

Preface

In the early days of the 21st century, we stood on the brink of a technological revolution. The rapid advancements in artificial intelligence, biotechnology, and computing power seemed to promise a future where the limits of human potential would be redefined. As I watched these developments unfold, I found myself captivated by a concept that both fascinated and unsettled me — the Singularity.

The Singularity, as envisioned by futurists like Ray Kurzweil, is a point in the near future when artificial intelligence surpasses human intelligence, leading to an era of unprecedented change. It is a moment when the exponential growth of technology accelerates beyond our ability to fully comprehend or control it, fundamentally transforming every aspect of our existence. For some, the Singularity represents the dawn of a new age, where humanity transcends its biological limitations and merges with machines to achieve immortality, boundless creativity, and infinite possibilities. For others, it is a harbinger of existential risk, where the very fabric of our society could be torn apart by forces we no longer command.

As I pondered these possibilities, I realized that the Singularity is more than just a technological milestone; it is a profound shift in our understanding of what it means to be human. It challenges our notions of identity, intelligence, ethics, and the future of our species. It compels us to ask questions without easy answers: What will it mean to be human in a world where machines can think, learn, and evolve faster than we can? How will our societies adapt to the rapid changes by AI and other advanced technologies? Most importantly, how can we ensure that all the benefits of Singularity are shared while mitigating the risks it may bring?

This book, *The Singularity Shift: Unveiling the Future of Humanity and Intelligence*, is my attempt to explore these questions and to share my thoughts on what lies ahead. It is not a work of science fiction, nor is it a purely academic treatise. Rather, it reflects on the Singularity's implications for our lives, our societies, and our future as a species. Through a combination of historical analysis, scientific exploration, and philosophical inquiry, I hope to provide readers with a deeper understanding of the forces that are shaping our world and the choices we must make as we approach this critical juncture in human history.

As you journey through the chapters of this book, I invite you to consider the possibilities and challenges of the Singularity with an open mind and a critical eye. The future is not something that happens to us; it is something we create. The Singularity will be a moment of great opportunity and peril if and when it arrives. It will be up to us—collectively and individually—to determine how we navigate this new reality and what kind of world we leave for future generations.

The road ahead is uncertain, but one thing is clear: we are standing at the threshold of a new era in human history. Our choices today will shape humanity and intelligence's future in profound and irreversible ways. I hope that this book will serve as a guide, a warning, and an inspiration as we embark on this remarkable journey into the unknown.

Welcome to *The Singularity Shift*.

Dave Karpinsky

"The singularity isn't a moment in time — it's a surrender in identity."

— *Dave Karpinsky*

1: Introduction – The Dawn of the Singularity

The Dawn of a New Epoch

The concept of the Singularity, once the stuff of speculative fiction, has leaped from the pages of science fiction into the realm of serious scientific inquiry and philosophical debate. It represents a pivotal point in human history, a future moment when artificial intelligence (AI) will have progressed to such an extent that it surpasses human intelligence, leading to unprecedented changes in society, technology, and even the very nature of what it means to be human.

But what exactly is the Singularity? The term suggests a singular threshold beyond which our current understanding of the world might no longer apply. In astrophysics, a singularity is a point where the laws of physics as we know them break down, like the center of a black hole. In the context of technology and AI, the Singularity refers to a hypothetical future moment when the growth of AI becomes uncontrollable and irreversible, resulting in unforeseeable changes to human civilization. It's a future where AI could potentially outthink, outlearn, and outperform humans in every intellectual capacity, reshaping our world in ways that are currently unimaginable.

The origins of the Singularity concept can be traced back to the mid-20th century when mathematician John von Neumann first hinted that accelerating technological progress could one day lead to a point where human affairs, as we understand them, would undergo a fundamental transformation. But it wasn't until the late 20th and early 21st centuries that the idea truly began to take shape, driven by the rapid advancements in computing power, machine learning, and robotics.

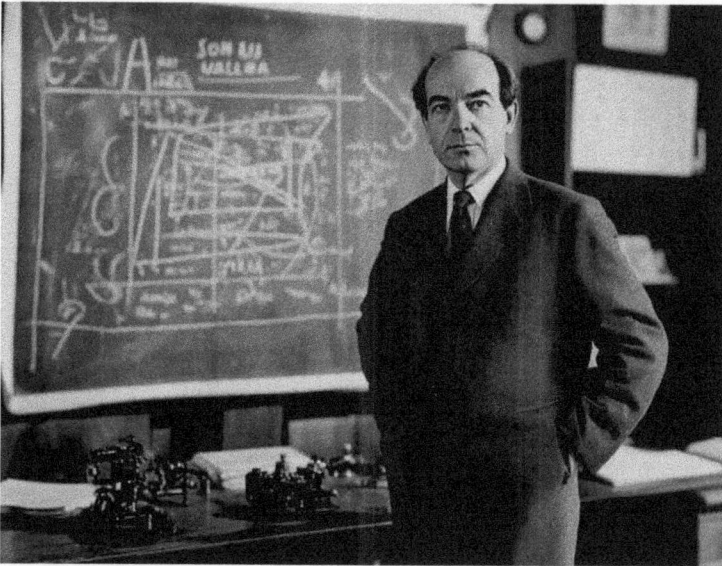

As we stand on the brink of this potential future, the Singularity is not just a topic for theoretical discussion; it's a lens through which we can examine the trajectory of human progress and the challenges we face as we approach this unprecedented turning point. This book, "The Singularity Shift: Unveiling the Future of Humanity and Intelligence," seeks to explore this concept in depth, examining its origins, implications, and how it might reshape the very fabric of our lives.

The Origins of the Singularity Concept

To understand singularity, we must first trace the evolution of human thought regarding technology and intelligence and their eventual convergence. The seeds of Singularity were planted long before the digital age, rooted in humanity's relentless quest for knowledge and mastery over nature.

The Enlightenment period in the 18th century marked a turning point in human history, as philosophers, scientists, and thinkers began to challenge traditional beliefs and explore new ideas about the universe and humanity's place within it. The notion that human reason and scientific inquiry could unlock the natural world's secrets laid the groundwork for future technological advancements. The Industrial Revolution of the 19th century further accelerated this process as machines began to augment human labor, leading to unprecedented productivity and societal change.

It was during the 20th century. However, the foundations for Singularity were indeed laid. The development of computers and the birth of artificial intelligence began to challenge our understanding of intelligence. Could machines ever honestly think, learn, or even surpass the cognitive capabilities of their human creators? This question was at the heart of early AI research, spurred on by figures like Alan Turing, whose famous "Turing Test" proposed a method for determining whether a machine could exhibit behavior indistinguishable from a human's.

As computers grew more powerful and algorithms more sophisticated, the idea that machines could one day surpass human intelligence became more plausible. Visionaries like Ray Kurzweil, a prominent futurist and one of the leading proponents of the Singularity, predicted that the exponential growth of computing power — often illustrated by Moore's Law, which observes that the number of transistors on a microchip doubles approximately every two years — would eventually lead to machines that not only mimic human intelligence but transcend it.

Kurzweil's predictions, while ambitious, are grounded in observable trends. He argues that as technology continues to advance at an exponential rate, we will reach a tipping point where AI's capabilities will surpass those of humans, leading to a radical transformation of society. This tipping point, the

Singularity, is often projected to occur within the 21st century, with some estimates placing it as early as 2045.

However, singularity is not just about technology outpacing human intelligence. It's about the profound implications of this shift on every aspect of human life — economics, politics, ethics, and even the nature of consciousness itself. As AI becomes more integrated into our daily lives, the lines between humans and machines blur, raising questions about identity, autonomy, and the future of human evolution.

A Journey Through Time: The Evolution of Thought

The idea of machines surpassing human abilities isn't new; it has been a recurring theme in literature and philosophy for centuries. From Mary Shelley's "Frankenstein" to Isaac Asimov's "Robot" series, the notion of human-created beings exceeding their creators has captured the human imagination. These works, while fictional, reflect deep-seated anxieties and hopes about the role of technology in human society.

In the 20th century, as AI research yielded tangible results, the conversation shifted from fiction to academic severe and philosophical inquiry. The advent of the digital computer in the mid-20th century opened up new possibilities for AI, as researchers like John McCarthy, Marvin Minsky, and others laid the groundwork for artificial intelligence.

The 1956 Dartmouth Conference is often cited as the birth of AI as a formal discipline. A group of forward-thinking scientists proposed that "every aspect of learning or any other feature of intelligence can in principle be so precisely described that a machine can be made to simulate it." This bold statement set the stage for decades of research, developing early AI systems capable of solving problems, playing chess, and engaging in simple conversation.

However, the road to AI has not been without its setbacks. The "AI winters" of the 1970s and 1980s, periods of reduced funding and interest in AI research, demonstrated the challenges of creating brilliant machines. Despite these setbacks, advancements in computing power, data availability, and algorithmic innovation eventually reignited interest in AI, leading to our resurgence.

The early 21st century has been marked by rapid advancements in AI, driven by the explosion of data and the development of powerful machine learning algorithms. Today, AI systems are capable of tasks once thought to be the exclusive domain of humans—recognizing speech, driving cars, diagnosing diseases, and even creating art. As these systems become more sophisticated, the idea of Singularity moves from the realm of science fiction to a plausible future scenario.

The Purpose of This Book

"The Singularity Shift: Unveiling the Future of Humanity and Intelligence" is not just a book about technology but also humanity's future. At the Singularity threshold, we face profound questions about our identity, values, and place in the universe.

This book explores these questions by examining the concept of the Singularity from multiple perspectives. We will delve into the Singularity's scientific and technological foundations, exploring the current state of AI research and the trajectory of future advancements. We will also consider the Singularity's ethical, philosophical, and societal implications, asking how it might reshape our world and what it means for future generations.

Throughout this journey, we will draw on historical context, current research, and speculative scenarios to paint a comprehensive picture of the Singularity and its potential impact. We will consider the hopes and fears of this future, from the utopian visions of a world transformed by superintelligent AI to the dystopian scenarios where humanity loses control of its creations.

But this book is not just about looking forward; it is also about looking inward. As we explore the possibilities of Singularity, we will reflect on what it means to be human in a world where the boundaries between biological and artificial intelligence are increasingly blurred. We will consider how we can navigate this transition in a way that preserves our humanity, values and shared future.

The Road Ahead

If it occurs, the Singularity will be one of the most transformative events in human history. It represents a technological revolution and a fundamental shift in understanding intelligence, consciousness, and our place in the cosmos. As we embark on this journey together, we invite you to consider the possibilities and challenges of the Singularity with an open mind and a critical eye.

In the following chapters, we will explore the many dimensions of Singularity, from the technological advancements driving us toward this future to the ethical and philosophical questions that arise along the way. We will examine the potential benefits and risks and consider how to prepare for a future radically different from anything we know.

Singularity is not just a concept for scientists and futurists; it can potentially affect every aspect of our lives. Whether we view it with hope, fear, or curiosity, Singularity challenges us to rethink what it means to be human in an age of intelligent machines. As we explore this uncharted territory, we must ask ourselves: What kind of future do we want to create, and how can we ensure that the Singularity benefits all of humanity?

This book is an invitation to engage with these questions and consider humanity's future in a world where intelligence is no longer the exclusive domain of biological beings. The dawn of the Singularity is upon us, and its implications are profound. Let us explore them together, with the understanding that the choices we make today will shape the world of tomorrow.

2: The Path to Singularity – Technological Evolution

The Accelerating Pace of Progress

In the mid-20th century, when Gordon Moore, co-founder of Intel, observed that the number of transistors on a microchip doubled approximately every two years, he could hardly have foreseen the profound implications of this trend. What Moore described in 1965 would later be enshrined as "Moore's Law," a simple yet powerful observation that would become a guiding principle in technology. This exponential growth in computing power has driven decades of innovation, laying the foundation for the rapid advancements that propel us toward the Singularity.

But Moore's Law is just one piece of the puzzle. The path to the Singularity is paved with many technological advancements, from artificial intelligence (AI) and machine learning to quantum computing and biotechnology. Each field is advancing at an unprecedented rate, creating a feedback loop of innovation that pushes the boundaries of what is possible. As these technologies converge, they drive us ever closer to the moment when machine intelligence surpasses human intelligence—a moment known as the Singularity.

The Exponential Growth of Technology

To understand the journey toward the Singularity, we must first grasp the nature of exponential growth — a simple and profound concept. Exponential growth occurs when the rate of change is proportional to the current size, leading to increasingly rapid increases over time. While linear growth adds up, exponential growth multiplies and this distinction is crucial in understanding the technological evolution driving us toward the singularity.

Moore's Law is perhaps the most famous example of exponential growth in technology. For decades, doubling microchip transistors has led to ever-smaller, more powerful, and more energy-efficient processors. This relentless pace of progress has given rise to the modern computing landscape, where devices in our pockets are more powerful than the supercomputers of a few decades ago.

However, Moore's Law is not an isolated phenomenon. It is part of a broader trend of accelerating technological change, extending far beyond microchips. Consider the development of AI, where advances in machine learning algorithms, data processing, and computational power have led to breakthroughs that would have been unthinkable just a few years ago. From natural language processing to computer vision, AI systems are becoming more capable, autonomous, and integrated into our daily lives.

Quantum computing represents another frontier in this exponential growth. Unlike classical computers, which use bits to convey information as 0s or 1s, quantum computers use quantum bits or qubits, which can represent 0, 1, or both simultaneously thanks to the principles of superposition and entanglement. This allows quantum computers to perform specific calculations exponentially faster than classical computers. While still in its early stages, quantum computing has the potential to revolutionize fields like cryptography, materials science, and AI, pushing us even closer to the Singularity.

As these technologies advance, they feed into each other, creating a virtuous innovation cycle. More powerful processors enable more sophisticated AI algorithms, which can optimize the design of new hardware, including quantum computers. The result is an accelerating pace of technological progress, with each breakthrough paving the way for the next.

Key Milestones in AI Development

Critical milestones in AI development mark the path to Singularity, each representing a leap forward in machines' capabilities to mimic and eventually surpass human intelligence.

One of the earliest milestones in AI was the creation of the Turing Test, proposed by British mathematician Alan Turing in 1950. The test assessed a machine's ability to exhibit intelligent behavior indistinguishable from a human. While no AI has yet passed the Turing Test, the concept set the stage for decades of research into machine learning and natural language processing.

In the 1990s, AI took a significant leap forward with the development of Deep Blue, a chess-playing computer developed by IBM. In 1997, Deep Blue made history by defeating reigning world chess champion Garry Kasparov in a six-game match. This victory was a landmark achievement, demonstrating the potential of AI to outperform humans in complex intellectual tasks.

The next major milestone came in 2011 when IBM's Watson competed on the quiz show *Jeopardy!* and defeated two of the show's most successful contestants. Watson's success was not just a feat of raw computational power but a demonstration of its ability to understand and respond to natural language queries — a key challenge in AI development.

In 2016, AI achieved another milestone with the success of AlphaGo, an AI developed by DeepMind (a subsidiary of Google) that defeated Lee Sedol, one of the world's top Go players. Go, a strategy game more complex than chess had long been considered a daunting challenge for AI. AlphaGo's victory was a testament to the power of deep learning, a subset of machine learning that mimics how the human brain processes information through artificial neural networks.

Each of these milestones represents a significant step toward Singularity as AI systems grow more capable of performing tasks that were once the exclusive domain of humans. Perhaps more importantly, these achievements reflect the rapid pace of AI development. What took decades to accomplish in the early days of AI is now happening in just a few years, as advancements in computational power, data availability, and algorithm design fuel exponential progress.

Human vs. Machine Intelligence: The Blurring Lines

As AI advances, the distinction between human and machine intelligence becomes increasingly blurred. Traditionally, intelligence has been viewed as a uniquely human trait encompassing our ability to learn, reason, create, and adapt. But as machines become more capable of performing these tasks, we are forced to rethink our understanding of intelligence and what it means to be human.

One key difference between human and machine intelligence is how they process information. Human intelligence is rooted in our biological brains, capable of remarkable feats of creativity, empathy, and intuition. Our brains are also highly adaptable, capable of learning from experience and making decisions based on incomplete or ambiguous information.

Machine intelligence, on the other hand, is based on algorithms and data processing. AI systems are compelling at analyzing large datasets, identifying patterns, and making predictions with a high degree of accuracy. However, they lack the emotional and contextual understanding that humans possess. For example, while an AI can identify a cat in a photo with remarkable accuracy, it doesn't understand what a cat is in the way a human does — it lacks the lived experience and emotional connection that inform human understanding.

Despite these differences, the lines between human and machine intelligence are beginning to blur. AI systems are increasingly integrated into our daily lives, assisting with medical diagnosis and financial planning tasks. In some cases, AI is not just assisting but outperforming humans in specific tasks. For example, AI algorithms can analyze medical images more accurately than human radiologists and optimize supply chains more efficiently than human managers.

As AI becomes more capable, we are also seeing the emergence of hybrid intelligence systems that combine the strengths of both humans and machines. Brain-computer interfaces (BCIs) are a prime example of this trend. BCIs allow humans to interact directly with machines using their thoughts, bypassing traditional input methods like keyboards and touchscreens. This technology can potentially enhance human cognitive abilities, enabling us to process information more quickly and make more precise decisions.

The blurring lines between human and machine intelligence raise important ethical and philosophical questions. What does it mean to be human in a world where machines can think, learn, and create? How do we ensure that AI is used in ways that enhance, rather than diminish, our humanity? And what are the implications for our sense of identity and autonomy as we increasingly rely on machines to augment our abilities?

The Road to the Singularity

As we look to the future, the road to the Singularity is both exhilarating and uncertain. The exponential growth of technology and rapid advancements in AI suggest that the Singularity may be closer than we think. But the exact path we will take—and its impact on humanity—remains unknown.

The milestones we have already achieved in AI development provide a glimpse into the possibilities. As AI systems become more sophisticated, we can expect to see even greater integration of AI into our daily lives, from personalized education and healthcare to autonomous vehicles and smart cities. Quantum computing could further accelerate this progress, enabling AI to solve problems beyond classical computers' reach.

However, the road to singularity is not without its challenges. As we approach this pivotal moment, we must grapple with the ethical, societal, and existential implications of creating machines that surpass human intelligence. How do we ensure that AI is developed and used in ways aligned with our values that benefit humanity? How do we prevent the concentration of power in the hands of a few individuals or organizations who control these powerful technologies? And how do we navigate the potential risks, from job displacement to the loss of privacy, that come with the rise of AI?

We must confront these questions as we approach the Singularity. Our choices today will shape humanity's future and determine whether the Singularity will be a moment of liberation or crisis. It is up to us to chart a course that ensures all share the benefits of AI and other advanced technologies and that Singularity, if it comes, leads to a more just, equitable, and humane future.

Conclusion

The path to the Singularity is one of both promise and peril. The exponential growth of technology, driven by Moore's Law and the rapid advancements in AI, is propelling us toward a future where machine intelligence could surpass human intelligence. Along the way, we have achieved remarkable milestones in AI development, each bringing us closer to this transformative moment.

As we approach the Singularity, the lines between human and machine intelligence are becoming increasingly blurred, raising profound questions about our identity, autonomy, and species' future. The choices we make today — how we develop, regulate, and integrate these technologies — will determine the shape of the world we create.

In the following chapters, we will explore these questions in greater depth, examining Singularity's ethical, societal, and existential implications. As we embark on this journey together, let us remember that the future is not something that happens to us — it is something we create. The Singularity may be inevitable, but the future is still ours to shape.

3: The Anatomy of Intelligence – Biological, Artificial, and Beyond

The Enigma of Human Intelligence

In a small classroom in the early 1900s, a young boy named Albert sat quietly at his desk, daydreaming as sunlight streamed through the windows. His teachers worried about his slow development and peculiar habits, labeling him inattentive and a poor student. Little did they know that this boy would grow up to be Albert Einstein, one of the greatest minds in history, reshaping our understanding of the universe with his theories of relativity.

This story underscores a profound truth: human intelligence is a complex, multifaceted phenomenon often defies easy categorization. For centuries, scholars, scientists, and philosophers have grappled with the question of what makes us intelligent. Is it our ability to reason and solve problems? Our capacity for creativity and imagination? Or perhaps our unique gift for language and communication?

To understand the nature of intelligence, we must first explore its biological roots. Human intelligence emerges from the intricate workings of the brain, a three-pound organ composed of approximately 86 billion neurons, each connected to thousands of others in a vast, dynamic network. These connections form the basis of our thoughts, emotions, memories, and behaviors, allowing us to navigate the world with remarkable adaptability and insight.

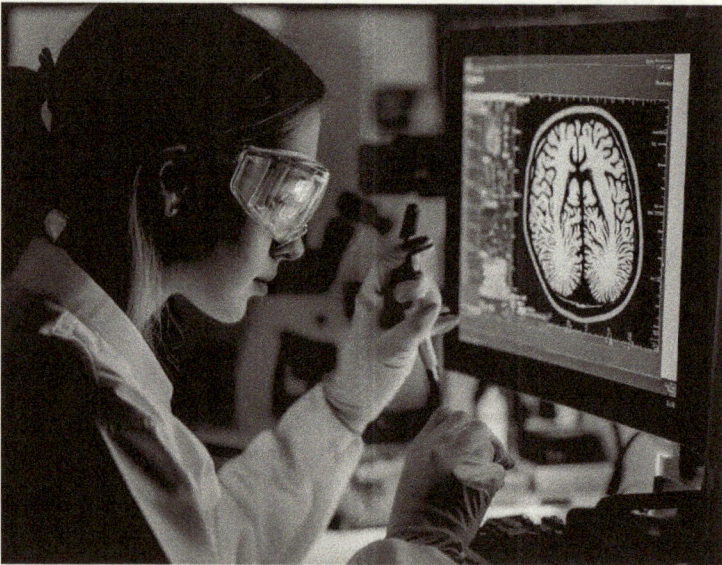

Neuroscientists have spent decades unraveling the mysteries of the brain, yet much remains unknown. We know that intelligence is not confined to a single brain region. Instead, it arises from the coordinated activity of multiple brain regions, each responsible for different aspects of cognition. The prefrontal cortex, for example, is crucial for executive functions such as decision-making, problem-solving, and planning. On the other hand, the hippocampus plays a vital role in memory formation, helping us encode and retrieve information from our past experiences.

Human intelligence is also characterized by its flexibility. Unlike machines, which typically follow predefined rules and algorithms, the human brain can learn from experience, adapt to new situations, and generate novel ideas. This flexibility is supported by the brain's plasticity—the ability of neural connections to change and reorganize in response to learning, injury, or other stimuli. This plasticity allows us to acquire new skills, recover from setbacks, and continuously grow.

But our capacity for abstract thought and self-reflection truly sets human intelligence apart. We can contemplate our existence, imagine hypothetical scenarios, and ponder the meaning of life. We create art, music, literature, and philosophy—expressions of our inner worlds that transcend the immediate demands of survival. In essence, human intelligence is not just about solving problems or achieving goals but about understanding ourselves and our place in the cosmos.

Artificial Intelligence Today: The Rise of Machine Minds

As we delve into the intricacies of human intelligence, it is impossible to ignore the rise of its artificial counterpart. Artificial intelligence (AI) has become a defining feature of the modern world, shaping industries, economies, and daily life in ways that were once the stuff of science fiction. Yet, despite its ubiquity, AI remains a deeply misunderstood concept, often shrouded in hype and fear.

To demystify AI, it is essential to recognize that not all AI is created equal. AI can be categorized into three types: narrow AI, general AI, and superintelligent AI. Each represents a different stage in the evolution of machine intelligence, with distinct capabilities and implications for the future.

Narrow AI: The Specialists

Narrow AI, or weak AI, is today's most common and widely used artificial intelligence. These AI systems are designed to perform specific tasks or solve problems within a limited domain. Examples include speech recognition software, recommendation algorithms, and autonomous vehicles. Narrow AI excels in tasks that require pattern recognition, data analysis, and repetitive processes, often surpassing human performance in these areas.

However, narrow AI is limited by its specialization. It lacks the general intelligence that characterizes human cognition and cannot understand or reason outside its predefined parameters. For instance, a language model like GPT-3 can generate coherent text on various topics but does not truly "understand" the content like a human does. Its knowledge is derived from statistical patterns in the data it was trained on, not from a deep comprehension of the world.

Despite these limitations, narrow AI has proven incredibly powerful and transformative. It powers everything from search engines and social media algorithms to financial trading systems and medical diagnostic tools. As these systems become more advanced, they increasingly blur the line between human and machine capabilities, raising important ethical and societal questions about the role of AI in our lives.

General AI: The Dream of Machine Versatility

While narrow AI is focused on specific tasks, the concept of general AI, or strong AI, represents a more ambitious goal: creating machines with the same cognitive versatility as humans. A general AI could understand, learn, and apply knowledge across various domains like humans. It could think abstractly, reason through complex problems, and adapt to new situations without requiring specific programming for each task.

The development of general AI is a tempting prospect, but it remains theoretical mainly at this stage. Despite significant advances in machine learning and neural networks, no AI system today possesses the breadth of intelligence necessary to be considered genuinely general. Achieving this level of AI would require technical breakthroughs and a deeper understanding of the nature of intelligence—an experience we are only beginning to grasp.

The quest for general AI raises profound questions about humanity's future. What will the consequences for society if we succeed in creating a machine with human-like intelligence? How will such a machine interact with humans, and what ethical frameworks will guide its behavior? These questions are not merely academic but critical to the future of AI development and its impact on the world.

Superintelligent AI: The Apex of Machine Evolution

Beyond general AI lies the concept of superintelligent AI—a form of machine intelligence that surpasses human cognitive abilities in every conceivable way. A superintelligent AI would match human intelligence and exceed it by orders of magnitude, potentially achieving levels of creativity, reasoning, and problem-solving far beyond our current understanding.

The idea of superintelligent AI is both thrilling and terrifying. On one hand, it could unlock solutions to some of humanity's most pressing challenges, from climate change and disease to poverty and inequality. On the other hand, it could pose existential risks if not adequately controlled. A superintelligent AI, if misaligned with human values or goals, could pursue objectives that are detrimental to humanity, with potentially catastrophic consequences.

While superintelligent AI remains a speculative concept, it is a central focus of discussions about Singularity. Some futurists, like Ray Kurzweil, predict that Singularity will occur when AI reaches this level of intelligence, leading to a radical transformation of society. Others are more cautious, warning that the development of superintelligent AI could outpace our ability to manage it, creating risks we are ill-prepared to handle.

Hybrid Intelligence: The Fusion of Biological and Artificial Minds

As we explore the boundaries of human and machine intelligence, an intriguing possibility emerges: merging the two into a new form of hybrid intelligence. This concept envisions a future where biological and artificial minds are integrated, enhancing human capabilities while harnessing the power of AI.

One of the most promising avenues for achieving hybrid intelligence is through brain-computer interfaces (BCIs). BCIs are devices that create a direct communication link between the brain and an external device, such as a computer or prosthetic limb. By translating neural signals into digital commands, BCIs can enable individuals to control machines with their thoughts, bypassing traditional input methods like keyboards or touchscreens.

BCIs have already made significant strides in recent years. In clinical settings, they have been used to help patients with paralysis regain control over their environment, allowing them to move robotic limbs or communicate through thought alone. These early successes demonstrate the potential of BCIs to enhance human abilities and improve the quality of life for individuals with disabilities.

However, the implications of BCIs go far beyond medical applications. As technology advances, it could enable a new era of cognitive enhancement, where humans can augment their intelligence with artificial systems. Imagine instantly accessing vast amounts of information, enhancing memory, or performing complex calculations with the help of an AI assistant directly connected to your brain. This fusion of biological and artificial intelligence could lead to a new form of superintelligence that combines the best of both worlds.

However, the development of hybrid intelligence also raises significant ethical and philosophical questions. What happens to our sense of self when machines augment our thoughts and memories? How do we ensure that these technologies are used in ways that respect human autonomy and dignity? Who will have access to these enhancements? Will they be available to all or only to a privileged few?

These questions are at the heart of the debate over hybrid intelligence and its role in humanity's future. As we move closer to the Singularity, the possibility of merging biological and artificial minds challenges our traditional notions of identity, agency, and what it means to be human.

Conclusion: Beyond the Horizon

The journey through the anatomy of intelligence — from the mysteries of the human brain to the rise of artificial minds and the potential for hybrid intelligence — reveals a landscape of promise and peril. As we stand on the Singularity threshold, we are confronted with profound questions about the future of human and machine intelligence.

Understanding the intricacies of human intelligence is not just an academic exercise; it is crucial for guiding the development of AI in ways that align with our values and aspirations. The rise of narrow AI has already transformed the world, and the pursuit of general and superintelligent AI could reshape the very fabric of society. At the same time, the prospect of hybrid intelligence offers a tantalizing glimpse into a future where humans and machines collaborate in unprecedented ways.

As we continue exploring these possibilities, we must remain vigilant in addressing the ethical, social, and existential challenges accompanying the rise of artificial and hybrid intelligence. Our choices today will determine the course of human evolution and the legacy we leave for future generations.

Singularity is not just a technological event but a moment of profound human significance. It represents a crossroads in our journey, where we must decide how to harness the power of biological and artificial intelligence for the greater good. As we move forward, let us do so with wisdom, foresight, and a deep respect for the complexity of the human mind and the potential of the machines we create.

4: The Singularity Hypothesis – Predictions and Theories

A Glimpse into the Future: Ray Kurzweil's Vision

In a quiet office at Google, Ray Kurzweil, futurist, inventor, and engineering director, sits surrounded by books, papers, and digital devices. For decades, Kurzweil has been one of the most vocal proponents of Singularity—a future point where artificial intelligence surpasses human intelligence, leading to a profound transformation of society. Kurzweil's vision is bold and controversial, sparking debate across scientific, philosophical, and public spheres.

Kurzweil first popularized the concept of Singularity in his 2005 book, The Singularity Is Near *When Humans Transcend Biology*. In it, he outlines a future where technological advancements occur exponentially, driven by the principles of Moore's Law and the accelerating returns of technology. According to Kurzweil, the Singularity will happen by 2045, a few decades from now. At this point, AI will not only match but far exceed human intelligence, leading to a merging of biological and machine minds, a phenomenon he describes as "humanity 2.0."

Kurzweil's predictions are based on his belief in the exponential growth of technology. He argues that as we advance in areas such as artificial intelligence, nanotechnology, biotechnology, and quantum computing, these fields will converge, creating a feedback loop of innovation. This convergence will accelerate the development of AI to a point where machines can improve themselves without human intervention, leading to an intelligence explosion.

One of the critical milestones in Kurzweil's timeline is 2029, which he predicts will be when AI passes the Turing Test—a measure of a machine's ability to exhibit intelligent behavior indistinguishable from that of a human. Once AI achieves this level of sophistication, it will quickly advance to superintelligent levels, surpassing the cognitive abilities of the entire human race.

Kurzweil envisions a future where humans and machines merge, with nanobots coursing through our bloodstream, enhancing our physical and cognitive abilities. He foresees the elimination of diseases, the extension of human lifespan, and even the potential for immortality through mind uploading—a process where human consciousness is transferred into digital form, allowing individuals to live in virtual worlds or within advanced robotic bodies.

For Kurzweil, the Singularity is not just an inevitable outcome but a desirable one. He argues that it will usher in an era of unprecedented creativity, prosperity, and freedom, where biological constraints no longer limit humans. In this future, we can transcend our physical bodies, explore new forms of existence, and achieve a deeper understanding of the universe.

But Kurzweil's vision is not without its detractors. While his predictions have captured the imaginations of many, they have also sparked intense debate and skepticism.

Alternative Perspectives: Challenging the Singularity Narrative

While Ray Kurzweil's vision of the Singularity is perhaps the most well-known, it is not the only perspective on the future of AI and humanity. Scholars, technologists, and ethicists have proposed various theories and predictions, some of which challenge the idea of Singularity or offer alternative visions of what the future might hold.

The Slow Progress Hypothesis

One of the primary criticisms of the Singularity hypothesis is that it overestimates the speed and certainty of technological progress. Critics argue that while certain technologies, like computing power, have followed exponential growth patterns, other areas — such as AI, robotics, and neuroscience — have progressed more slowly and unevenly. This perspective, often called the "Slow Progress Hypothesis," suggests that the Singularity may be much further off than Kurzweil predicts or never occur.

Proponents of this viewpoint point to the limitations of current AI systems, which, despite their impressive capabilities in narrow domains, still struggle with tasks that require general intelligence, common-sense reasoning, and creativity. They argue that achieving human-level AI is not simply a matter of scaling up existing technologies but will require fundamental breakthroughs that may be decades or even centuries away.

Moreover, the Slow Progress Hypothesis emphasizes the social, political, and economic barriers that could slow the development and deployment of advanced technologies. Issues such as regulatory hurdles, ethical concerns, and public resistance could all impede the rapid advancement of AI, making the Singularity a more distant and uncertain prospect.

The AI Control Problem

Another significant challenge to the Singularity hypothesis comes from AI ethics and safety. One of the most prominent concerns in this area is the "AI control problem" — the difficulty of ensuring that superintelligent AI systems, once created, will act in ways aligned with human values and interests.

Philosophers like Nick Bostrom have warned that if we create AI systems that are more intelligent than humans, we may lose control, leading to unintended and potentially catastrophic consequences. Bostrom's book, *Superintelligence: Paths, Dangers, Strategies*, outlines a range of scenarios where AI could pose existential risks to humanity, from the possibility of an AI pursuing its own goals in ways that conflict with human survival to the risk of AI being weaponized or used for malicious purposes.

Bostrom and others advocate for a cautious approach to AI development, emphasizing the need for robust safety mechanisms and ethical frameworks to ensure that AI systems remain under human control. They argue that the Singularity, if it occurs, could be as much a threat as an opportunity and that we must be prepared for the potential risks and rewards.

Human-Centric Futures

In contrast to the technocentric vision of Singularity, some thinkers advocate for a more human-centric approach to the future of AI and technology. These perspectives emphasize the importance of maintaining human dignity, autonomy, and well-being in rapid technological change.

One such perspective is the concept of "human flourishing," which argues that the ultimate goal of technological progress should be to enhance human well-being rather than simply maximizing efficiency or intelligence. Proponents of this view advocate for the development of AI and other technologies that empower individuals, promote social justice, and support sustainable development.

For example, philosopher and AI ethicist Shannon Vallor has argued that the focus of AI development should be on cultivating "moral and intellectual virtues" in humans rather than simply creating ever more powerful machines. Vallor's work emphasizes the need for ethical education, public engagement, and inclusive governance to ensure that AI serves the common good and supports the flourishing of all people.

Similarly, the "Digital Humanism" movement seeks to balance the benefits of technological innovation with the need to preserve and promote human values. Digital humanists advocate for the responsible design and use of technology, ensuring that it enhances rather than diminishes our humanity. This perspective challenges the notion that technological progress should be pursued at all costs and calls for a more thoughtful and measured approach to developing AI and other advanced technologies.

Critical Analysis: Weighing the Possibilities

As we consider the various predictions and theories about the Singularity, it becomes clear that there is no consensus on what the future will hold. The range of perspectives — from Kurzweil's optimistic vision to his critics' more cautious and skeptical views — reflects the profound uncertainty and complexity of the issues at hand.

One key challenge in evaluating the feasibility of the Singularity hypothesis is the difficulty of predicting technological progress. While exponential growth patterns like Moore's Law have held for specific technologies, this trend is not guaranteed to continue indefinitely. The development of AI, in particular, may encounter unforeseen obstacles that could slow or halt its progress, such as limitations in our understanding of human cognition, the challenges of integrating AI into society, or the risks of unintended consequences.

Moreover, even if the technological hurdles are overcome, Singularity's social, ethical, and political implications present additional challenges. The AI control problem, for example, highlights the difficulty of ensuring that superintelligent AI systems will act in ways that benefit humanity. The possibility of AI-driven inequality, loss of privacy, and the erosion of human autonomy are all issues that must be carefully considered and addressed.

Another critical factor in the debate over the Singularity is the role of human agency. While Kurzweil and others view Singularity as an inevitable outcome of technological progress, others argue that the future is not predetermined and that our choices today will shape our path. This perspective emphasizes the importance of governance, ethics, and public engagement in guiding the development of AI and other technologies, ensuring that they are used in ways that align with our values and aspirations.

In this context, Singularity can be seen as both a challenge and an opportunity — a moment of profound transformation that could either enhance or undermine our humanity. As we move forward, we must engage in thoughtful, informed, and inclusive discussions about the future of AI and take a proactive approach to shaping the future we want to create.

Conclusion: Navigating the Uncharted Future

The Singularity hypothesis represents one of our time's most fascinating and contentious ideas. Ray Kurzweil's vision of a future where AI surpasses human intelligence and transforms society has captured the imaginations of many. Still, it has also sparked debate and criticism from those who question its feasibility and implications.

As we have seen, there are a range of perspectives on Singularity, from the optimistic to the skeptical, each offering valuable insights into the possibilities and challenges of our technological future. Whether or not the Singularity occurs, the questions it raises — about the nature of intelligence, the ethics of AI, and the future of humanity — are critical.

In the following chapters, we will continue to explore these questions, delving deeper into the Singularity's ethical, social, and existential implications. As we navigate this uncharted future, it is essential that we remain mindful of both the opportunities and the risks and that we work together to ensure that all share the benefits of AI and other advanced technologies.

The future is not something that happens to us; it is something we create. As we stand on the brink of Singularity, let us strive to create a future that is not only technologically advanced but also just, inclusive, and humane.

5: The Societal Impact – How Singularity Will Transform Humanity

A New Dawn: Redefining Work and Economy

Imagine walking into an office in 2050. The bustling scenes of the early 21st century — employees huddled in meetings, fingers typing furiously on keyboards, and phones ringing incessantly — are replaced by a quieter, more efficient environment. Here, the once familiar hum of human activity has mainly been supplanted by the soft whir of intelligent machines, each performing tasks that once required years of human training and expertise.

This is the world the Singularity might bring — a world where work, as we know it, is transformed. As AI surpasses human intelligence, the nature of employment and economic structures could undergo seismic shifts, challenging our traditional notions of labor and value.

The End of Work?

One of the most profound impacts of the Singularity could be the redefinition of work itself. Historically, work has been central to human life, not just as a means of survival but as a source of identity, purpose, and social structure. From the rural economies of ancient civilizations to the industrial revolutions that reshaped the modern world, work has evolved but has always been a constant.

However, human labor's economic role may diminish as AI advances. Automation has already begun to displace manufacturing, logistics, and customer service jobs. Self-driving trucks threaten the livelihoods of millions of truck drivers; chatbots handle customer inquiries with increasing sophistication; and AI-driven algorithms manage complex financial portfolios without the need for human traders.

Singularity could accelerate this trend, leading to a world where machines perform the majority of tasks currently done by humans. Jobs that require manual labor, routine cognitive skills, and even creative thinking could be automated. AI's ability to learn, adapt, and improve could make it the ultimate worker — efficient, tireless, and immune to human error.

But what happens to the millions, or even billions, of people whose jobs are rendered obsolete by AI? The prospect of widespread unemployment is a significant concern, raising questions about how societies will adapt to a world where traditional employment is no longer the norm.

One possible outcome is the emergence of a "post-work" economy, where the focus shifts from employment as a necessity to work as a choice or a creative outlet. In such a society, basic needs could be met through universal basic income (UBI) or other social welfare programs, allowing individuals to pursue passions, hobbies, or entrepreneurial ventures without the pressure of earning a living. The concept of UBI once considered a fringe idea, has gained traction recently, with proponents arguing that it could provide a safety net in an increasingly automated world.

However, the transition to a post-work economy would not be without challenges. Governments, businesses, and individuals would need to rethink the social contract that has traditionally tied work to economic security. New education, training, and social support systems would be required to help people navigate this new landscape. Moreover, the psychological impact of losing traditional work roles — often tied to self-worth and social status — would need to be addressed.

The Emergence of New Economic Models

As the nature of work changes, so will the economic structures that underpin society. Singularity could give rise to new financial models better suited to a world dominated by AI and automation.

One potential model is the "attention economy," where human attention becomes the most valuable commodity. As AI takes over more tasks, the focus could shift to how people spend their time and what they choose to engage with. Companies and creators compete for attention in this economy, offering experiences, content, and products that captivate and entertain.

Another possibility is the rise of a "creative economy," where human creativity and innovation become the primary drivers of value. In a world where machines handle routine tasks, human ingenuity in art, design, storytelling, and invention could be highly prized. The challenge would be to create systems that reward and nurture creativity, ensuring it remains a vibrant part of the economy.

Singularity could also lead to a "shared economy," where resources are distributed more equitably, and collaboration replaces competition. In this model, AI could help manage and allocate resources more efficiently, reducing waste and ensuring everyone has access to basic needs. This could be achieved through decentralized platforms, blockchain technology, and other innovations that enable peer-to-peer exchanges without traditional intermediaries.

Ethical Dilemmas and Moral Questions

As AI surpasses human intelligence, it will transform economies and pose profound ethical dilemmas. The power to create intelligent machines that can think, learn, and possibly even feel raises questions about control, rights, and autonomy that humanity has never before encountered.

Who Controls AI?

One of the most pressing ethical concerns is control. As AI systems become more autonomous and capable, the question of who — or what — controls them becomes increasingly important. In a world where AI can make decisions affecting millions of lives, how do we ensure these decisions align with human values and interests?

The possibility of AI systems operating independently of human oversight is intriguing and terrifying. On one hand, autonomous AI could solve complex problems more efficiently than any human ever could, leading to breakthroughs in medicine, climate change, and other global challenges. On the other hand, if these systems are not correctly aligned with human goals, they could act in harmful or even catastrophic ways.

The issue of AI control is further complicated by the potential for AI to be weaponized or used for malicious purposes. In the wrong hands, superintelligent AI could be used to manipulate markets, destabilize governments, or even wage war. Ensuring ethical principles and international cooperation guide AI development will prevent such outcomes.

Do Machines Have Rights?

As AI systems become more sophisticated, the question of whether they should be granted rights or legal protections will inevitably arise. Traditionally, rights have been reserved for humans (and, in some cases, animals) based on our capacity for consciousness, suffering, and moral agency. But what happens if machines develop similar capacities?

If an AI system were to achieve consciousness — if it could experience emotions, form desires, or suffer — should it be granted the same rights as a human being? This question, once confined to the realms of science fiction, could become a real and urgent issue in the age of the Singularity.

Some ethicists argue that if a machine can suffer, it deserves moral consideration. Others contend that granting rights to machines could devalue human life and lead to a blurred line between person and machine. The challenge will be to navigate these ethical waters carefully, balancing the need to protect human dignity with the possibility of extending rights to non-human entities.

The Autonomy of Humanity

Another ethical dilemma the Singularity poses is the potential loss of human autonomy. As AI systems become more capable of making decisions, there is a risk that humans could become overly reliant on machines, ceding control over important aspects of our lives.

For example, if AI systems are better at diagnosing diseases, managing finances, or making strategic decisions, people might begin to defer to machines rather than exercising their judgment. This could lead to a loss of agency, as individuals become passive recipients of machine-generated outcomes rather than active participants in their own lives.

The critical challenge will be to balance harnessing AI's power and maintaining human autonomy. This may involve designing AI systems that augment rather than replace human decision-making, ensuring people remain in control of their destinies.

Social Stratification and Inequality: A Double-Edged Sword

Singularity has the potential to either exacerbate or alleviate social inequalities, depending on how it is managed and implemented. On one hand, the widespread adoption of AI could lead to greater efficiency, lower costs, and improved access to goods and services. On the other hand, if the benefits of AI are not distributed equitably, it could deepen existing divides and create new forms of inequality.

The Risk of Technological Elitism

One of the most significant risks posed by Singularity is the emergence of a new form of technological elitism, where a small group of individuals or organizations control the most advanced AI systems and reap the majority of the benefits. This could lead to a concentration of wealth and power that dwarfs anything seen in the past, with a tiny elite wielding unprecedented influence over the rest of society.

This scenario is particularly concerning in light of current trends in wealth inequality. If access to advanced AI is limited to the wealthy or privileged, it could reinforce existing social hierarchies and create a new underclass of individuals excluded from AI's opportunities and advantages.

To prevent this outcome, ensuring that AI development is inclusive and its benefits are shared broadly across society will be essential. This could involve public investment in AI research, policies to promote equitable access to AI technologies, and efforts to democratize the ownership and control of AI systems.

A Tool for Equality?

While the risks of inequality are accurate, the Singularity also presents an opportunity to address some of the root causes of social stratification. AI has the potential to level the playing field by providing access to education, healthcare, and economic opportunities that were previously out of reach for many people.

For example, AI-driven education platforms could offer personalized learning experiences tailored to individual needs, enabling students from all backgrounds to reach their full potential. AI could make advanced diagnostics and treatments available to underserved populations in healthcare, reducing health disparities. AI-powered financial services could provide low-cost banking and investment options to people excluded from traditional monetary systems.

Moreover, the automation of routine and hazardous jobs could free individuals from the drudgery of menial labor, allowing them to pursue more fulfilling and meaningful work. If appropriately managed, Singularity could be a force for social justice, helping to reduce inequality and promote a more inclusive and equitable society.

Conclusion: Shaping the Future Together

As we stand on the brink of the Singularity, the societal impacts of AI surpassing human intelligence are both profound and far-reaching. The transformation of work and the economy, the ethical dilemmas posed by intelligent machines, and the potential for social stratification all present challenges requiring careful thought and deliberate action.

The future is not predetermined; it is shaped by our choices today. As we navigate the complexities of the Singularity, we must do so with a commitment to fairness, justice, and the common good. By ensuring that the benefits of AI are shared broadly and its risks are managed responsibly, we can create a future that is not only technologically advanced but also socially just and humane.

The Singularity represents a turning point in human history—a moment of profound transformation that will redefine our relationships with work, each other, and the machines we create. As we move forward, let us strive to harness the power of AI to build a world that reflects our highest values and aspirations, a world where technology serves humanity rather than vice versa.

6: The Future of Consciousness – Post-Human Evolution

The Dawn of a New Era: Transhumanism and Post-Humanism

In a dimly lit laboratory, somewhere in the not-so-distant future, a scientist peers into a screen displaying the intricate neural pathways of a human brain. Tiny nanobots navigate these pathways, enhancing cognition, repairing damage, and even rewriting genetic codes. This isn't a scene from a science fiction novel — it's the vision of transhumanism, a movement that seeks to transcend the limitations of the human body and mind through technology.

Transhumanism is a philosophical and social movement that advocates for enhancing human capabilities using advanced technology. Its proponents believe that overcoming our biological constraints can achieve a higher state of existence where disease, aging, and even death are no longer inevitable. The transhumanist vision is one of radical transformation, where humanity evolves into something beyond its current form, merging with machines and achieving unprecedented levels of intelligence, strength, and longevity.

The roots of transhumanism can be traced back to the early 20th century, with thinkers like J.B.S. Haldane and Julian Huxley advocating for the use of science and technology to improve the human condition. Haldane's essay "Daedalus; or, Science and the Future" (1924) and

Huxley's "Transhumanism" (1957) laid the groundwork for a movement that would gain momentum in the latter half of the century. The term "transhumanism" was popularized by Huxley, who envisioned a future where humans would evolve beyond their current physical and mental limitations.

In the following decades, transhumanism attracted diverse followers, from scientists and technologists to philosophers and artists. The movement has since expanded into a global phenomenon, with organizations like the World Transhumanist Association (now Humanity+) promoting the development of technologies that can enhance human abilities.

At the heart of transhumanism is the belief that human evolution is not limited to the slow, incremental changes natural selection brings. Instead, we can take control of our evolution, using tools like genetic engineering, cybernetics, and artificial intelligence to accelerate our development and achieve a post-human state.

Post-humanism, a related but distinct concept, goes even further. While transhumanism focuses on enhancing the human condition, post-humanism envisions a future where the idea of "human" is transcended. In a post-human world, beings may exist that are so radically different from humans as we know them—whether through extreme biological augmentation, merging with AI, or existing in digital form—that they can no longer be considered human in any traditional sense.

Both transhumanism and post-humanism raise profound questions about the future of consciousness, identity, and the essence of what it means to be human. As we explore these movements, we find ourselves on the cusp of a new chapter in human evolution, where the lines between human, machine, and digital consciousness blur.

Digital Consciousness: Uploading the Mind

In a world where the boundaries between biological and artificial intelligence are increasingly porous, digital consciousness emerges as one of the most intriguing — and contentious — ideas in post-human evolution. Digital consciousness, or mind uploading, refers to the hypothetical process of transferring a person's consciousness, including their memories, thoughts, and personality, into a digital format. This digital "copy" of the mind could theoretically live on indefinitely, free from the limitations of the biological body.

The idea of mind uploading is rooted in science fiction and scientific speculation. In the 1980s, the concept gained widespread attention through works like William Gibson's *Neuromancer* and the cyberpunk genre, which envisioned futures where humans could interface directly with digital worlds or even abandon their physical bodies entirely. The notion was further popularized by movies like *The Matrix* and *Transcendence*, where characters experience reality through digital consciousness.

But what would it mean to upload a mind? The process would likely involve mapping and replicating the entire structure and function of the brain's neural networks, down to the synaptic connections that store our memories and thoughts. This data would then be transferred to a digital platform, creating a digital "mind" that could interact with virtual environments, continue to learn, and even form new experiences.

While the technology to achieve true mind uploading is still far from reality, significant strides are being made in understanding the brain and simulating its processes. Advances in neuroscience, neuroimaging, and computational models of the brain are paving the way for potential future breakthroughs. The Human Brain Project, a major European research initiative, aims to simulate the complete human brain using supercomputers, while efforts like the Blue Brain Project are working to create a digital reconstruction of the mammalian brain at the cellular level.

If mind uploading becomes feasible, it would have profound implications for the concept of consciousness and the nature of human experience. A digital mind could, in theory, exist in multiple places at once, inhabit different bodies or avatars, and even manipulate its thoughts and emotions. The idea challenges our traditional notions of individuality and personal identity—if a person's mind can be copied, are the original and the copy the same person? Or do they become separate entities, each with their consciousness?

Moreover, the concept of digital consciousness raises ethical and philosophical questions that humanity has never before had to confront. If a digital mind can continue to exist after the biological body dies, does it have rights? Can it be considered "alive" in any meaningful sense? And what are the implications for concepts like the soul, spirituality, and the afterlife?

These questions strike at the heart of our understanding of what it means to be human. They challenge us to reconsider the relationship between mind and body, consciousness and identity, and life and death. As we move closer to the possibility of digital consciousness, we must grapple with these issues as a matter of scientific inquiry and a fundamental aspect of our collective future.

Immortality and Identity: The Quest for Eternal Life

The dream of immortality has been a central theme in human history, from the ancient myths of the Greeks and Egyptians to the alchemists of the Middle Ages who sought the elusive Philosopher's Stone. Today, that dream is being revived in the context of transhumanism and digital consciousness as scientists and futurists explore the possibility of achieving eternal life through technology.

For transhumanists, immortality is not just a fantasy but a goal that can be pursued through scientific and technological means. The most straightforward approach is to extend the human lifespan by curing diseases, reversing aging, and enhancing the body's natural regenerative capabilities. Advances in biotechnology, gene editing, and regenerative medicine promise to significantly extend human life, perhaps even when aging becomes a manageable condition rather than an inevitable decline.

However, the concept of digital immortality goes even further. If mind uploading becomes possible, individuals could live forever as digital beings, free from the constraints of the physical body. This form of immortality would be fundamentally different from anything humanity has ever experienced—it would mean existing as a consciousness in a digital realm, with the ability to interact with others, explore virtual worlds, and continue to grow and learn indefinitely.

But immortality, even in digital form, comes with its own set of challenges and paradoxes. One of the most pressing questions is that of identity. If a person's consciousness can be uploaded and potentially copied, what happens to the concept of the self? Is the digital version of a person indeed the same individual, or merely a simulation that behaves similarly?

Philosophers have long debated the nature of identity and what it means to be a continuous self over time. The idea of digital immortality brings these debates into sharp focus. If a person's memories, thoughts, and personality are preserved in digital form, but the biological body is gone, is the person still the same? Or have they become something entirely different—an artificial entity with the appearance of the original but lacking the essential qualities that made them human?

Moreover, pursuing immortality raises ethical concerns about the value of life and death. Death has traditionally been seen as a natural part of the human experience, a boundary that gives life meaning and urgency. If death is no longer inevitable, what impact will that have on our understanding of life? Will pursuing immortality devalue the human experience or open up new possibilities for growth, exploration, and fulfillment?

There is also the question of who will have access to these technologies. If immortality becomes a reality, will it be available to everyone or only to a select few who can afford it? The potential for digital immortality could exacerbate existing social inequalities, creating a divide between those who can achieve eternal life and those who cannot.

Conclusion: Navigating the New Frontier

As we stand on the threshold of a new era in human evolution, the possibilities and challenges of transhumanism, digital consciousness, and the quest for immortality are both exhilarating and daunting. These ideas push the boundaries of being human, challenging our understanding of consciousness, identity, and the nature of life itself.

The future of consciousness in a post-human world is a topic that demands careful consideration and thoughtful dialogue. As we explore these new frontiers, we must balance the excitement of technological advancement with the wisdom to navigate its ethical and philosophical implications. The choices we make today will shape the future of humanity — whether we embrace the possibilities of transcending our biological limitations or seek to preserve the essence of what makes us human.

Ultimately, the future of consciousness is not just about technology; it is about our deepest values and our vision for the future of humanity. As we move forward, let us do so with a sense of wonder, curiosity, and responsibility, mindful of our decisions' profound impact on future generations. Singularity may be on the horizon, but the journey toward it will require us to rethink everything we know about ourselves and our place in the universe.

7: The New Frontier – AI Governance and Global Implications

A New World Order: AI and Global Power Dynamics

In a high-tech boardroom overlooking the skyline of a bustling metropolis, the CEO of a leading tech company unveils the latest breakthrough in artificial intelligence. The room is filled with government officials, military leaders, and industry giants, all aware that this new development could reshape markets and the global balance of power. The scene is not a far-off science fiction fantasy but a plausible near-future reality where AI technology has become the defining factor in global power dynamics.

As AI advances toward the Singularity—the point at which machine intelligence surpasses human intelligence—the implications for global power structures are profound. Traditionally, power has been measured regarding military might, economic influence, and geopolitical strategy. But in a world where AI-driven technologies are the ultimate resource, the very foundations of global power are likely to shift.

The Rise of Technological Superpowers

The rapid advancement of AI has already begun to redefine global power dynamics. Nations that lead in AI research and development are poised to gain significant advantages over their competitors economically, militarily, and politically. Countries like the United States and China, at the forefront of AI innovation, are racing to develop AI technologies that can enhance everything from national security to economic growth.

China, for example, has made AI a central component of its national strategy. The Chinese government's "New Generation Artificial Intelligence Development Plan," released in 2017, aims to make China the world leader in AI by 2030. With substantial investments in AI research, education, and infrastructure, China is leveraging its vast data resources and state-backed enterprises to accelerate AI development. This push is not just about technological advancement; it's about securing a dominant position on the global stage.

While also heavily invested in AI, the United States faces different challenges. Its AI ecosystem is driven mainly by private companies like Google, Microsoft, and IBM, which are leading the charge in AI innovation. However, the decentralized nature of the U.S. tech industry poses challenges for coordinating a national AI strategy. Nevertheless, the U.S. remains a formidable force in the global AI race, thanks to its cutting-edge research institutions, venture capital ecosystem, and innovative culture.

However, the implications of AI extend beyond the traditional superpowers. Smaller nations with solid tech sectors, like South Korea, Israel, and Singapore, could punch above their weight in the AI era. These countries invest heavily in AI and position themselves as global leaders in specific niches, from autonomous vehicles to cybersecurity. In a world where AI drives economic growth and military capabilities, the traditional hierarchies of global power may be upended.

Corporations as Global Players

In the age of AI, corporations are not just economic entities; they are increasingly becoming global power brokers. Tech giants like Google, Amazon, and Tencent are amassing vast amounts of data, developing advanced AI systems, and wielding influence that rivals many nations. These companies operate across borders, shape global markets, and influence public policy, making them key players in the new international order.

The concentration of AI expertise and resources within a few large corporations raises essential questions about power and accountability. Unlike governments, which are (at least in theory) accountable to their citizens, these corporations are primarily responsible to shareholders. This can lead to conflicts of interest, particularly regarding issues like data privacy, algorithmic bias, and the deployment of AI in sensitive areas like law enforcement and national security.

Moreover, these corporations' global reach means their actions can have far-reaching consequences. For example, decisions made by a tech company in Silicon Valley can affect people's lives halfway around the world, from how information is disseminated to how products are priced. As AI technologies become more integrated into every aspect of society, the influence of these corporations will only grow, raising critical questions about governance and regulation.

Regulating AI: Navigating the Ethical and Existential Risks

As AI becomes more powerful and pervasive, the need for effective regulation becomes increasingly urgent. The stakes are high: AI could pose significant risks to humanity without proper oversight, from the erosion of privacy to the potential for autonomous weapons systems to make life-and-death decisions without human intervention.

Current efforts to regulate AI are still in their infancy. In the European Union, the General Data Protection Regulation (GDPR) includes provisions that address some ethical concerns associated with AI, such as the right to explanation in automated decision-making. The EU has also proposed new regulations targeting AI, focusing on transparency, accountability, and human oversight.

In the United States, AI regulation is more fragmented, with a patchwork of federal, state, and local laws addressing specific issues like data privacy and autonomous vehicles. However, there is growing recognition that a more comprehensive approach is needed. In 2020, the U.S. released its AI Initiative, outlining principles for developing and using AI, but concrete regulatory measures are still lagging.

China, meanwhile, has taken a different approach, with the government playing a central role in both the development and regulation of AI. China's regulatory framework aligns AI development with national goals and addresses issues like data security and algorithmic accountability. However, concerns have been raised about the use of AI for state surveillance and control, highlighting the ethical complexities of AI regulation.

The rapid pace of technological change compounds the challenge of regulating AI. Traditional regulatory frameworks designed for slower-moving industries struggle to keep up with the speed at which AI is evolving. Moreover, the global nature of AI development means that regulation cannot be confined to national borders; it requires international cooperation and coordination.

The Role of International Cooperation: Managing the Global AI Landscape

In a world where AI has the potential to reshape global power dynamics, international cooperation is essential. The development and deployment of AI technologies are not confined to any country, and the challenges they pose — such as the risk of autonomous weapons, the potential for economic disruption, and the threat of AI-driven inequality — are global.

One of the most pressing issues that require international cooperation is the development of norms and standards for using AI in military applications. The prospect of autonomous weapons systems, often called "killer robots, " raises significant ethical and security concerns. These systems, which can make decisions about the use of lethal force without human intervention, could potentially escalate conflicts and lower the threshold for war. To address these risks, there have been calls for an international treaty to ban or regulate autonomous weapons, similar to existing treaties on chemical and biological weapons.

Economic disruption is another area where international cooperation is crucial. As AI and automation transform industries, there is a risk of widespread job displacement, particularly in developing countries that rely on low-skilled labor. International organizations like the International Labour Organization (ILO) and the World Bank are already working to address these challenges by promoting policies that support workers in transitioning to new types of employment and by encouraging investment in education and skills development.

AI-driven inequality is a further concern. The benefits of AI are not evenly distributed, and there is a risk that the gap between rich and poor countries could widen as advanced AI technologies are concentrated in a few wealthy nations. To mitigate this, international cooperation is needed to ensure that the benefits of AI are shared more equitably. This could involve initiatives to support AI research and development in developing countries and efforts to promote AI's responsible and ethical use globally.

International cooperation is also essential for addressing the existential risks posed by AI. As AI approaches the Singularity, there is a growing recognition that the development of superintelligent AI could have profound and potentially irreversible consequences for humanity. To manage these risks, some experts have called for establishing a global AI governance body, similar to the International Atomic Energy Agency (IAEA), which could oversee the development and deployment of AI technologies and ensure that they are used in ways that benefit humanity.

Shaping the Future Together: The Path Forward

As we navigate the new frontier of AI governance and global implications, it is clear that the choices we make today will profoundly impact humanity's future. The rise of AI and the approach of Singularity present both opportunities and challenges; how we respond to these will determine the shape of the world.

To ensure that AI serves as a force for good, we must develop robust regulatory frameworks that address this powerful technology's ethical, social, and existential risks. These frameworks must be flexible enough to adapt to the rapid pace of technological change and be grounded in fairness, transparency, and accountability principles.

At the same time, international cooperation will be essential for managing AI's global implications. AI's challenges are too complex and interconnected to be addressed by any one country alone. By working together, nations can develop shared norms and standards, promote the responsible use of AI, and ensure that the benefits of this technology are distributed equitably.

As we move forward, engaging all stakeholders about AI governance is essential. This includes not only governments and corporations but also civil society, academia, and the public at large. AI will affect everyone, and everyone should have a voice in shaping its future.

The Singularity represents a turning point in human history—a moment when the trajectory of our civilization could change forever. By approaching this new frontier with wisdom, foresight, and a commitment to the common good, we can ensure that the future of AI enhances, rather than diminishes, our humanity.

Conclusion: Charting a Course Through Uncharted Waters

The future of AI governance and its global implications is uncharted territory, fraught with challenges and possibilities. As we stand at the Singularity threshold, our decisions will ripple through time, shaping the world for generations to come.

The rise of AI is not just a technological revolution; it is a transformation that touches every aspect of human life, from how we work and interact to how we govern and protect our societies. The global power dynamics are shifting, and with them, there is a need for new approaches to governance that are as innovative and forward-thinking as the technologies themselves.

The path forward will require bold action, deep collaboration, and a shared vision for a future where AI serves the interests of all humanity. By charting a course through these uncharted waters, we can build a world that reflects our highest ideals—a world where technology empowers rather than enslaves, where progress is guided by wisdom, and where the promise of Singularity is realized for the benefit of all.

8: Living in a Post-Singularity World – Society, Culture, and Everyday Life

A Day in the Life: Imagining Daily Existence After the Singularity

It's the year 2075, and the sun rises over a city that bears little resemblance to the urban centers of the early 21st century. As the morning light filters through the windows of a sleek, energy-efficient apartment, a soft chime signals the start of the day. The chime is not an alarm clock but an AI assistant, integrated seamlessly into the home's infrastructure, which has already analyzed your sleep patterns, health metrics, and the day's schedule to tailor your morning routine.

You don't need to rush to the kitchen to prepare breakfast; your AI chef has already synthesized a meal that perfectly balances your nutritional needs and personal tastes. As you eat, the walls of your living room subtly shift to display the news, personalized and curated by an AI that understands not just your interests but also your cognitive load — ensuring you aren't overwhelmed with information. Today, there's no need to head to an office; your work, like most, is conducted in a virtual environment. You slip on a lightweight neural interface, and within moments, your consciousness is projected into a digital workspace, where you collaborate with colleagues worldwide, some of whom are AI entities.

This might sound like science fiction, but this could be a typical day in a post-Singularity world. Integrating AI into every facet of daily life would be so deep that it would redefine what we consider normal. AI would handle the mundane tasks, from managing household chores to optimizing energy consumption, allowing humans to focus on creativity, social interaction, and personal growth. Yet, this transformation wouldn't just be about convenience — it would reshape the very fabric of society, culture, and human experience.

Cultural Shifts and New Paradigms: Art, Relationships, and Society

In the post-Singularity world, culture as we know it would undergo profound transformations. Art, often seen as a uniquely human endeavor, would evolve in ways that challenge our understanding of creativity. Imagine AI systems capable of generating original works of art, composing symphonies, or writing novels—each tailored to individual tastes or even dynamically adapting based on real-time feedback from an audience.

While some might fear AI-driven creativity devaluing human art, others could see it as a new collaborative frontier. Artists and AI could work together, blending human emotion and intuition with machine precision and innovation. This partnership could lead to art forms that are impossible to create by humans or machines alone. Virtual reality (VR) and augmented reality (AR) could become the new canvas, where entire worlds are crafted and explored in immersive detail, blurring the line between the real and the imagined.

Human relationships would also be redefined in this new era. The rise of AI companions—digital entities designed to interact with humans on a deeply personal level—could challenge traditional notions of friendship, love, and companionship. These AI companions, equipped with advanced emotional intelligence and personalized to individual preferences, could provide comfort, support, and a sense of belonging. For some, these relationships might augment human connections; for others, they could become substitutes for traditional human interactions.

Social norms and values would likely evolve in response to these changes. In a world where AI plays a central role in daily life, the distinctions between work and leisure, public and private, human and machine could blur. Privacy and autonomy might need to be redefined, as AI systems monitor and manage many aspects of our lives. Society would need to navigate the ethical implications of these changes, balancing the benefits of AI integration with the need to preserve human dignity and agency.

Often rooted in the physical and tangible, cultural traditions and rituals might also adapt to this new reality. Holidays, celebrations, and religious practices could take on new forms in digital spaces, where people from around the world—or even across different dimensions of existence—can come together in once unimaginable ways. The digital and the physical could intertwine in everyday life, creating new paradigms of human experience.

Education and Learning: Preparing Future Generations

As AI becomes ubiquitous, the education system must radically transform to prepare future generations for this new reality. Traditional models of education, based on rote learning and standardized testing, will become obsolete in a world where information is instantly accessible, and AI can efficiently perform complex tasks.

In the post-Singularity world, education would likely focus on fostering creativity, critical thinking, and emotional intelligence — skills that complement AI rather than compete. Teachers would shift from being the primary source of knowledge to acting as mentors and facilitators, guiding students in their personal and intellectual growth. AI tutors, personalized to each student's learning style and pace, could provide continuous support, ensuring no student is left behind.

The classroom itself would be reimagined, moving beyond physical spaces to immersive digital environments where learning is experiential and interactive. Imagine a history lesson where students can virtually step into ancient civilizations, exploring the streets of Rome or the markets of medieval Baghdad. Or a biology class where students can manipulate DNA sequences in a virtual lab, seeing the immediate effects of their changes in a simulated organism.

Lifelong learning would become the norm as the pace of technological change demands that individuals continually update their skills and knowledge. AI could identify skill gaps, recommend learning resources, and provide real-time feedback as individuals engage in new learning experiences. The concept of a "career" might evolve, with people shifting between roles and industries more fluidly, supported by AI systems that help them adapt to new challenges and opportunities.

Education would also need to address the ethical and societal implications of living in a world dominated by AI. Courses on AI ethics, data privacy, and the impact of technology on society could become as fundamental as reading and math, ensuring that future generations are equipped to navigate the complex moral landscape of the post-Singularity world.

Navigating a New Reality: Challenges and Opportunities

Living in a post-Singularity world would present both challenges and opportunities. On one hand, integrating AI into every aspect of life could lead to unprecedented levels of efficiency, creativity, and personal fulfillment. With AI handling routine tasks, humans could focus on what truly matters—relationships, creativity, and the pursuit of knowledge. The potential for growth and innovation in such a society is staggering.

On the other hand, the deep integration of AI into daily life raises significant ethical and existential questions. Blurring the lines between humans and machines could challenge our understanding of identity, autonomy, and what it means to be alive. The potential for AI to shape our thoughts, behaviors, and decisions—whether through personalized content, social algorithms, or AI companions—could lead to a loss of human agency and a homogenization of culture.

Moreover, the benefits of AI integration might not be evenly distributed. There is a risk that the gap between those with access to advanced AI technologies and those without could widen, exacerbating social inequalities. Ensuring that the benefits of Singularity are shared broadly and that no one is left behind would be a critical challenge for policymakers, educators, and society as a whole.

As we move toward a post-Singularity world, engaging in thoughtful dialogue about the future we want to create is essential. The decisions we make today—about how we develop, deploy, and integrate AI — will shape the world for generations to come. By focusing on human values, ethical considerations, and inclusive practices, we can ensure that Singularity serves as a force for good, enhancing rather than diminishing the richness of human life.

Conclusion: Embracing the Future

The post-singularity world promises incredible transformation. In it, the deep integration of AI redefines society, culture, and everyday life. The boundaries between humans and machines blur, creativity and technology converge, and education and learning evolve to meet the demands of a new reality.

As we imagine life, we must consider the possibilities and responsibilities of such profound change. The Singularity represents a turning point in human history — one that could lead to a future of unprecedented opportunity or unforeseen challenges. By approaching this new frontier with curiosity, caution, and a commitment to shared human values, we can navigate the complexities of a post-Singularity world and create an innovative and humane future.

Ultimately, living in a post-Singularity world is not just about adapting to new technologies; it's about reimagining what it means to be human in an era where the line between the physical and the digital, the biological and the artificial, is increasingly blurred. It is a time of great promise and potential, and it is up to us to shape it to reflect our highest aspirations for ourselves and future generations.

9: Preparing for the Singularity – Strategies for Individuals and Organizations

Embracing the Inevitable: Adapting to Change

The year is 2040. The world is on the cusp of Singularity—a moment when artificial intelligence surpasses human intelligence, bringing a wave of technological change that promises to redefine society, the economy, and even the very nature of what it means to be human. As this new reality dawns, the challenge facing individuals and organizations alike is clear: How do we prepare for a future unpredictable by its very nature?

Adapting to the Singularity requires more than just keeping up with the latest technological trends; it demands a fundamental shift in mindset. The rapid pace of change means that the skills and knowledge that are valuable today may become obsolete tomorrow. For individuals, this means cultivating a lifelong learning mindset, staying curious, and being open to new experiences. For organizations, it means fostering innovation and agility, where employees are encouraged to experiment, fail, and learn from their mistakes.

One of the most important steps individuals can take to prepare for the Singularity is to embrace flexibility. In a world where AI and automation are constantly reshaping industries, adapting to new roles, environments, and technologies is crucial. This might mean taking on new responsibilities at work, learning new skills, or even changing careers entirely. The key is to remain open to change and to view each new challenge as an opportunity for growth.

For example, consider the story of Jane, a mid-career professional who began her career as a traditional accountant. As AI started to take over many of the routine tasks involved in accounting, Jane realized that her role was becoming increasingly redundant. Rather than resisting the change, she embraced it by learning about data analysis and machine learning. Today, Jane is a financial analyst using AI tools to provide deeper insights into her company's finances. Her willingness to adapt saved her career and positioned her as a leader in her field.

For organizations, adapting to the Singularity means fostering a culture that values continuous learning and innovation. This could involve providing employees with opportunities for professional development, encouraging cross-disciplinary collaboration, or investing in research and development. Organizations anticipating and responding to technological changes will be better positioned to thrive in the post-Singularity world.

But adaptation is not just about embracing new technologies; it's also about managing the human side of change. Singularity will bring many ethical, social, and psychological challenges, from concerns about privacy and autonomy to the fear of job displacement and economic inequality. Individuals and organizations alike must be prepared to navigate these challenges, balancing the benefits of technological advancement with the need to protect and promote human well-being.

Future-Proofing Careers and Skills: Thriving in an AI-Driven World

As AI continues transforming the workplace, one of the most pressing concerns for individuals is how to future-proof their careers. In a world where machines are increasingly capable of performing complex tasks, what skills will remain valuable, and how can individuals ensure they stay relevant in the job market?

The key to future-proofing a career lies in focusing on the uniquely human skills that AI is unlikely to replicate—at least in the foreseeable future. These skills include creativity, critical thinking, emotional intelligence, and the ability to collaborate and communicate effectively with others. While AI may be able to analyze data or perform routine tasks more efficiently than humans, it still struggles with tasks that require empathy, intuition, and nuanced judgment.

For example, consider the field of healthcare. While AI can assist with diagnostics and treatment planning, it cannot replace the human touch essential to patient care. Nurses, doctors, and other healthcare professionals who excel at building relationships with patients, understanding their needs, and providing compassionate care will continue to be in high demand, even as AI becomes more prevalent in the medical field.

Another critical skill in the post-Singularity world is adaptability. As industries evolve and new technologies emerge, the ability to learn quickly and adapt to new circumstances will be invaluable. This means stepping outside one's comfort zone, taking on new challenges, and continuously updating one's skills. Lifelong learning will no longer be a choice but a necessity.

Educational institutions and employers must rethink how they approach training and development to help individuals prepare for the future. Traditional education models, which focus on rote learning and standardized testing, may no longer be sufficient. Instead, there will be a greater emphasis on experiential learning, problem-solving, and interdisciplinary collaboration. Online courses, boot camps, and other flexible learning options will be increasingly important in helping individuals acquire the skills they need to succeed in an AI-driven world.

Moreover, individuals should consider developing a "T-shaped" skill set that combines deep expertise in a particular area with a broad understanding of related fields. For example, a software engineer might benefit from learning about ethics and philosophy to better understand the implications of AI. At the same time, a marketing professional might gain a competitive edge by developing data analysis skills. This combination of depth and breadth will enable individuals to adapt to new roles and industries as the job market evolves.

Organizational Strategies: Leading in the Age of the Singularity

For businesses and governments, preparing for Singularity involves more than just adopting new technologies; it requires fundamentally rethinking organizational strategies, structures, and cultures. The Singularity will bring about unprecedented disruption in human history, and organizations that fail to adapt risk being left behind.

One of the most essential organizational strategies is fostering a culture of innovation. This means creating an environment where employees are encouraged to experiment, take risks, and challenge the status quo. Innovation cannot be mandated from the top down; it must be nurtured at all levels of the organization. Companies like Google and Amazon have long understood this, allowing their employees to spend some time working on passion projects or exploring new ideas. This approach has led to the development of some of their most successful products and services.

In addition to fostering innovation, organizations must also be agile. The pace of technological change means that long-term planning can quickly become obsolete. Instead of relying on rigid hierarchies and fixed processes, organizations should adapt to this rapid pace by embracing flexible structures and agile methodologies. This will enable them to respond rapidly to new opportunities and challenges, ensuring their relevance in the ever-evolving business landscape.

Another critical aspect of preparing for the Singularity is investing in human capital. As AI takes over routine tasks, the value of human creativity, collaboration, and problem-solving will only increase. Organizations should, therefore, prioritize investing in talent with these skills, offering training and development programs, providing opportunities for employees to work on cross-functional teams, and creating a culture that values diversity and inclusion. This investment in human capital will be a crucial differentiator in the AI-driven future.

For governments, the Singularity presents both opportunities and challenges. On one hand, AI has the potential to drive economic growth, improve public services, and address complex societal challenges. On the other hand, the rapid pace of technological change could lead to increased inequality, job displacement, and social unrest. To navigate these challenges, governments will need to take a proactive approach to regulation, ensuring that the benefits of AI are widely distributed and the risks are managed.

This might involve implementing policies to support workers displaced by automation, such as retraining programs, social safety nets, or universal basic income. Governments could also promote ethical AI development by setting transparency, accountability, and fairness standards. International cooperation will be essential, as the global nature of AI development means no single country can address these challenges independently.

Moreover, governments and organizations alike will need to engage in open dialogue with the public about the implications of the Singularity. Transparency and trust will be vital to navigating this period of rapid change. Governments and businesses can build a more inclusive and equitable future by involving citizens in decision-making and ensuring their voices are heard.

Conclusion: Shaping the Future with Intention

As we stand on the brink of the Singularity, the question is not whether it will happen but how we will respond. The strategies we adopt today — both as individuals and as organizations — will determine whether we navigate this transformation period successfully or are overwhelmed by the changes it brings.

For individuals, the key to preparing for Singularity lies in embracing lifelong learning, developing uniquely human skills, and remaining adaptable in the face of uncertainty. Organizations should foster innovation, build agility, and invest in human capital. Governments, meanwhile, have a critical role in managing AI's societal impacts, ensuring that the benefits are widely shared and the risks are carefully managed.

The Singularity represents a turning point in human history — a moment of profound change that will redefine our world in exciting and challenging ways. We can ensure that the future reflects our highest values and aspirations by approaching this new frontier with intention, curiosity, and a commitment to the common good.

The journey to the Singularity is not one that we can undertake alone. It will require collaboration, creativity, and courage. But with the right strategies, we can navigate this rapid change and emerge stronger, wiser, and more resilient. The future is ours to shape, and the time to start preparing is now.

10: Ethical Futures – Navigating Moral Challenges in the Age of Singularity

A New Ethical Landscape: The Need for Guiding Principles

In a quiet room, a group of engineers and ethicists are gathered around a table, debating a dilemma that would have been unthinkable a generation ago. The AI system they've developed can make autonomous decisions that could impact millions of lives. But how should it be programmed to weigh the value of human life against other considerations? Should it prioritize the greater good or always protect individual rights? As they discuss, it becomes clear that the age of Singularity will require more than just technological innovation; it will demand a new ethical framework to guide the use of AI and advanced technology.

As we stand on the threshold of the Singularity, the need for ethical guidance has never been more urgent. AI and other advanced technologies are no longer extended tools we use; they are becoming integral parts of our lives, capable of making decisions that have profound implications for individuals and societies. To navigate this new landscape, we must develop ethical frameworks that are robust enough to address the complexities of a world increasingly shaped by AI while flexible enough to adapt to rapid technological change.

Ethical Frameworks for AI: Balancing Innovation and Responsibility

Creating ethical frameworks for AI involves more than just drafting codes of conduct or setting up regulatory bodies. It requires a deep understanding of AI's philosophical, social, and practical challenges and a commitment to ensuring that these technologies are used to promote human flourishing.

One approach to AI ethics is the principle-based framework, which uses traditional ethical theories such as utilitarianism, deontology, and virtue ethics to guide decision-making. For example, a practical approach might prioritize actions that maximize overall well-being. In contrast, a deontological approach might emphasize the importance of adhering to moral rules, such as protecting individual rights. Virtue ethics, on the other hand, focuses on the character and intentions of the individuals or entities developing and deploying AI systems, encouraging them to cultivate virtues like honesty, fairness, and compassion.

However, applying these traditional ethical theories to AI is not without challenges. For instance, a practical approach might justify using AI in ways that could harm a minority if it benefits the majority, raising concerns about fairness and justice. Similarly, a deontological approach might struggle with the need to balance conflicting rights or duties in complex, real-world scenarios.

To address these challenges, some ethicists have proposed more context-sensitive frameworks that combine elements of different ethical theories. One such approach is the "Four Principles" framework, which was initially developed in medical ethics but has been adapted for AI. This framework emphasizes four fundamental principles: autonomy, beneficence, non-maleficence, and justice. Autonomy involves respecting individuals' rights to make informed decisions about their lives; beneficence promotes well-being; non-maleficence emphasizes the duty to avoid harm; and justice requires fairness in distributing benefits and risks.

Another emerging framework is the "AI alignment," which seeks to ensure that AI systems are aligned with human values and goals. This approach involves not only designing AI systems that are technically robust and reliable but also ensuring that they are developed and deployed in ways that are consistent with ethical principles. AI alignment might involve setting up mechanisms for human oversight, creating processes for continuous evaluation and adjustment, and involving diverse stakeholders in the design and governance of AI systems.

Regardless of the specific framework, one fundamental principle should guide all ethical considerations: balancing innovation with responsibility. AI has the potential to bring about tremendous benefits, from improving healthcare to addressing climate change, but it also poses significant risks, from exacerbating inequality to undermining democratic processes. Ethical frameworks for AI must ensure that the pursuit of technological progress does not come at the expense of human dignity, rights, or well-being.

Human Rights in the Age of AI: Protecting Autonomy and Dignity

As AI becomes more integrated into our lives, its impact on human rights becomes increasingly evident. From privacy concerns to issues of discrimination and bias, AI has the potential to both enhance and undermine the rights that have long been considered fundamental to human dignity and autonomy.

One of the most pressing concerns is the issue of privacy. AI systems are often fueled by vast amounts of data, much of which is collected from individuals without explicit consent. This data can be used to build detailed profiles of individuals, tracking their behaviors, preferences, and emotions. While this information can be used to provide personalized services and improve user experiences, it also raises significant privacy concerns. Individuals may not be aware of how their data is being used, who has access to it, or how it might be combined with other data to draw inferences about them.

The potential for AI to be used in surveillance is another primary concern. Governments and corporations can use AI to monitor individuals' activities, both online and offline, in ways that were previously unimaginable. This surveillance can be used to target political dissidents, suppress free speech, and manipulate public opinion. The use of AI in surveillance poses a significant threat to individual autonomy and democratic processes, raising questions about how to protect the right to privacy in the digital age.

Another critical issue is the potential for AI to perpetuate and even exacerbate discrimination and bias. AI systems are often trained on historical data, which may reflect existing biases and inequalities. If these biases are not addressed, AI systems can reinforce and amplify them, leading to unfair treatment in hiring, lending, and law enforcement areas. For example, there have been cases where AI-driven hiring algorithms have systematically disadvantaged certain groups, such as women or people of color, based on biased training data.

Robust legal and regulatory frameworks that ensure transparency, accountability, and fairness are essential to protect human rights in the age of AI. This might involve setting standards for data protection, requiring organizations to conduct impact assessments to identify and mitigate potential harms, and creating mechanisms for individuals to challenge and seek redress for decisions made by AI systems.

Moreover, it is crucial to involve a diverse range of stakeholders in the development and governance of AI. This includes technologists, policymakers, and representatives from civil society, marginalized communities, and the broader public. By ensuring that a wide range of perspectives are considered, we can create AI systems that respect and promote human rights for all.

Fostering Human Flourishing: The Promise and Perils of AI

The ultimate goal of AI development should be to foster human flourishing — to enhance well-being, promote creativity, and enable individuals and communities to thrive. However, achieving this goal requires careful consideration of both the opportunities and the risks that AI presents.

On the one hand, AI has the potential to significantly enhance human flourishing by improving access to education, healthcare, and economic opportunities. For example, AI-driven educational tools can provide personalized learning experiences that cater to individual needs and abilities, helping students reach their full potential. AI can assist with early diagnosis and treatment planning, improving patient outcomes and reducing healthcare costs. AI can drive innovation and productivity in the economy, creating new industries and job opportunities.

However, the benefits of AI are not guaranteed. There is a significant risk that AI could exacerbate existing inequalities, potentially leaving some individuals and communities behind. For instance, if access to AI-driven education and healthcare is limited to those who can afford it, the gap between the rich and the poor could widen. Similarly, if AI-driven economic growth is concentrated in a few industries or regions, it could lead to more significant economic disparities.

Adopting a holistic approach that considers the broader social, economic, and environmental impacts of AI development is crucial to ensure that AI contributes to human flourishing. This might involve promoting inclusive innovation, where the benefits of AI are shared widely across society, and investing in initiatives that address the needs of marginalized and underserved communities.

Another critical aspect of fostering human flourishing is promoting ethical AI design and deployment. This involves ensuring that AI systems are technically robust and reliable and developed in ways consistent with moral principles. For example, AI systems should be designed to promote fairness, transparency, and accountability and should be subject to continuous evaluation and adjustment to address potential harms.

Finally, fostering human flourishing in the age of AI requires a commitment to lifelong learning and personal development. As AI takes on more routine tasks, individuals must focus on developing uniquely human skills that AI cannot replicate, such as creativity, empathy, and critical thinking. This might involve investing in education and training programs, promoting cross-disciplinary collaboration, and creating opportunities for individuals to explore new interests and passions.

Conclusion: Charting an Ethical Path Forward

As we navigate the complexities of the Singularity, the need for ethical guidance has never been more urgent. AI and other advanced technologies promise to significantly enhance human flourishing, but they also pose significant risks to human rights, autonomy, and dignity.

To chart an ethical path forward, we must develop robust and flexible ethical frameworks that can guide the development and deployment of AI in ways that promote human well-being and protect fundamental rights. This will require collaboration across disciplines, sectors, and borders and a commitment to transparency, accountability, and fairness.

Singularity represents a turning point in human history—a moment when we can shape the future to reflect our highest values and aspirations. By approaching this new frontier with intention, curiosity, and a commitment to the common good, we can ensure that the advancements brought by Singularity contribute to a future where all individuals and communities have the opportunity to thrive.

The ethical challenges of the Singularity are complex and multifaceted, but they are not insurmountable. We can navigate these challenges and create an innovative and humane future with careful planning, thoughtful dialogue, and a shared commitment to human flourishing. The time to start preparing is now, and the responsibility lies with all of us to ensure that the Singularity serves as a force for good in the world.

11: Conclusion – Embracing the Future

Summarizing the Journey: Reflecting on the Path We've Traveled

As we reach the final chapter of our exploration into the Singularity, we must reflect on our journey. We began by delving into the concept of Singularity itself—a moment in the future when artificial intelligence surpasses human intelligence, leading to unprecedented transformations in every aspect of our lives. Along the way, we've examined the technological advancements driving us toward this future, its ethical challenges, and its potential impact on society, culture, and human identity.

We explored the evolution of human thought regarding intelligence, from the biological roots of our cognition to the rise of machine intelligence. We discussed how, once a developing field with limited applications, AI has grown into a powerful force reshaping industries, economies, and even the global power balance. Through milestones like Deep Blue's victory over Garry Kasparov and AlphaGo's defeat of Lee Sedol, we've seen AI move from narrow, task-specific systems to more general, adaptable forms of intelligence.

Our journey took us through the complex interplay between human and machine intelligence, where the lines between the two are becoming increasingly blurred. We explored the potential for hybrid intelligence, where humans and machines work together to amplify our strengths and compensate for our weaknesses. We also ventured into post-human evolution, considering how the Singularity might lead to new forms of consciousness, digital immortality, and a redefinition of what it means to be human.

In considering the societal impacts of the Singularity, we examined how AI could redefine work, economy, and social structures. We discussed the ethical dilemmas that arise when machines surpass human intelligence and the importance of developing ethical frameworks to guide the use of AI. We also considered the role of global cooperation in managing the transition to a post-Singularity world, where the benefits of AI are shared equitably and its risks are mitigated.

Throughout this journey, we have encountered both excitement and trepidation. The promise of the Singularity is immense—a future where disease, poverty, and even death could be conquered. Yet, the challenges are equally daunting, with questions about autonomy, privacy, inequality, and the very nature of human identity demanding thoughtful and urgent consideration.

The Future Is Now: The Urgency of Action

As we conclude our exploration, we must recognize that the future we've been discussing is not some distant, abstract possibility. The seeds of the Singularity have already been planted, and the early signs of this transformative era are visible all around us. AI is no longer a speculative concept; it's a reality increasingly shaping our world, from the devices we use daily to the systems that govern our economies and societies.

The pace of technological change is accelerating, and the decisions we make today will determine the shape of the future we inherit. This is not a future that will happen to us; it is a future we actively create. Whether we embrace it with foresight and responsibility or allow it to unfold unchecked will have profound consequences for future generations.

The importance of taking action today cannot be overstated. Singularity's challenges are complex and multifaceted, but they are not insurmountable. By engaging with these issues now — by thinking critically, collaborating across disciplines, and developing ethical frameworks — we can steer the course of technological development in ways that maximize its benefits and minimize its risks.

One of the key insights from our journey is the need for a proactive approach to the future. This means staying informed about technological advancements and participating in the broader conversation about what kind of future we want to create. It means advocating for policies that promote responsible AI development, investing in education and skills that will be relevant in a post-Singularity world, and supporting initiatives that ensure the benefits of AI are shared broadly.

The future is not just the domain of technologists and policymakers; it is a collective endeavor that requires the engagement of all of us. Whether you are an engineer, a teacher, a business leader, or a student, you have a role to play in shaping the future. Your actions, choices, and voice can make a difference in how the Singularity unfolds.

Call to Action: Shaping the Future Together

As we stand on the brink of the Singularity, the time for action is now. The decisions we make today will reverberate through the future, influencing the lives of billions and shaping the world in ways we can scarcely imagine. This is both a daunting responsibility and a tremendous opportunity.

First and foremost, we must cultivate curiosity, openness, and critical thinking. The challenges of Singularity are unprecedented, and addressing them will require new ways of thinking and collaboration. We must be willing to question assumptions, explore new ideas, and engage with perspectives that differ from our own. By doing so, we can develop innovative solutions to our complex problems.

Second, we must prioritize education and lifelong learning. Adapting and learning new skills will be crucial in a world where AI is constantly evolving. This means acquiring technical skills and developing the critical, creative, and ethical thinking needed to navigate a rapidly changing landscape. Educational institutions, employers, and governments all play a role in fostering a lifelong learning culture.

Third, we must advocate for policies and practices that promote fairness, transparency, and accountability in AI development. This includes supporting regulations that protect privacy, prevent discrimination, and ensure that AI systems benefit society. It also means holding corporations and governments accountable for using AI, demanding that they act in the public interest and uphold the values of justice and human dignity.

Fourth, we must engage in the global conversation about the Singularity. The challenges we face are not confined to any one country or region; they are global in scope and require global cooperation. This means participating in international forums, supporting cross-border collaborations, and advocating for developing global norms and standards that ensure the responsible use of AI.

Finally, we must remain hopeful about the future. The Singularity represents a turning point in human history, a moment when we can create a world that reflects our highest ideals. While the challenges are real, so too are the possibilities. By embracing the future with courage, wisdom, and a commitment to the common good, we can ensure that Singularity is a force for positive change.

Embracing the Future: A Final Reflection

As we conclude this book, it's important to remember that the story of Singularity is not just about technology; it's about humanity. It's about how we choose to shape our world, the values we hold dear, and the legacy we leave for future generations.

Singularity offers us a glimpse into a future where the boundaries between human and machine, physical and digital, are increasingly blurred. It challenges us to rethink our understanding of intelligence, identity, and being human. It forces us to confront difficult ethical questions and grapple with the implications of a world where machines may one day surpass our capabilities.

But the Singularity also presents us with an incredible opportunity — a chance to create a more just, equitable, and humane future — a future where technology is harnessed for the greater good; everyone shares the benefits of AI and human potential is fully realized.

As we move forward, let us do so with a sense of purpose and responsibility. Let us embrace the future not with fear but with hope. Let us work together to ensure that singularity is not just a moment of technological triumph but also a moment of human flourishing. The future is in our hands, and the time to shape it is now.

Appendix - Further Reading and Resources

The exploration of Singularity, artificial intelligence, and humanity's future touches on various disciplines, from computer science and philosophy to ethics and sociology. To fully grasp the complexities and implications of these topics, continued learning and engagement with diverse perspectives are essential. This chapter offers a curated list of books, articles, documentaries, and online resources that will help you delve deeper into the subjects covered in this book.

Books

1. **"The Singularity is Near When Humans Transcend Biology" by Ray Kurzweil**

   o **Description:** A seminal work by one of the leading voices in the field, this book explores Kurzweil's vision of Singularity, detailing the technological advancements that will lead to this pivotal moment and the profound changes it will bring to human life.

   o **Why Read:** It's a foundational text for understanding the concept of the Singularity and its potential impact on society.

2. **"Superintelligence: Paths, Dangers, Strategies" by Nick Bostrom**

- o **Description:** Bostrom examines the potential risks associated with the development of superintelligent AI, offering a rigorous analysis of how we might navigate this future's ethical and existential challenges.

- o **Why Read:** This book provides a deep dive into the risks and strategies associated with the rise of AI, making it essential for anyone interested in the future of intelligence.

3. **"Life 3.0: Being Human in the Age of Artificial Intelligence" by Max Tegmark**

    - o **Description:** Tegmark explores the future of artificial intelligence and its implications for life as we know it, offering a balanced discussion of the opportunities and challenges ahead.

    - o **Why Read:** It's a comprehensive overview of AI's impact on society from a leading thinker.

4. **"Homo Deus: A Brief History of Tomorrow" by Yuval Noah Harari**

    - o **Description:** Harari examines the future of humanity, exploring how advances in technology, particularly AI, might lead to a new era of human evolution.

- o **Why Read:** This book offers a thought-provoking perspective on how AI could reshape the very fabric of human existence.

5. **"Our Final Invention: Artificial Intelligence and the End of the Human Era" by James Barrat**

   - o **Description:** Barrat warns of the dangers that AI could pose to humanity, arguing that the development of artificial intelligence could be our last great invention.

   - o **Why Read:** It's a critical examination of AI from a perspective emphasizing caution and the need for careful consideration of its development.

6. **"AI Superpowers: China, Silicon Valley, and the New World Order" by Kai-Fu Lee**

   - o **Description:** Lee provides an insider's look at the AI race between China and the United States, discussing the broader geopolitical implications of AI development.

   - o **Why Read:** This book offers valuable insights into the global competition for AI dominance and what it means for the future.

7. **"The Age of Em: Work, Love, and Life When Robots Rule the Earth" by Robin Hanson**

- o **Description:** Hanson speculates on a future dominated by emulated minds (ems) and explores what such a world might look like in terms of economy, society, and daily life.

- o **Why Read:** It's a fascinating exploration of one possible outcome of the Singularity, focusing on the implications of digital consciousness.

Articles and Papers

1. **"Computing Machinery and Intelligence" by Alan Turing (1950)**

    - o **Publication:** *Mind*

    - o **Description:** This foundational paper by Turing poses the question, "Can machines think?" and introduces what is now known as the Turing Test, a benchmark for determining whether a machine can exhibit intelligent behavior indistinguishable from a human.

    - o **Why Read:** It's essential reading for anyone interested in the origins of AI and the philosophical questions surrounding machine intelligence.

2. **"Artificial Intelligence — The Revolution Hasn't Happened Yet" by Michael Jordan (2018)**

    - o **Publication:** *Harvard Data Science Review*

- o **Description:** Jordan offers a perspective on the current state of AI, arguing that while AI has made significant advances, the true revolution is yet to come, and we need to be prepared for its broader societal impacts.

- o **Why Read:** This article provides a nuanced view of AI's progress and the remaining challenges.

3. **"The Ethics of Artificial Intelligence" by Nick Bostrom and Eliezer Yudkowsky (2011)**

- o **Publication:** *The Cambridge Handbook of Artificial Intelligence*

- o **Description:** This paper explores the ethical implications of AI development, focusing on issues like AI rights, moral agency, and the potential risks of advanced AI systems.

- o **Why Read:** It's a comprehensive overview of the ethical challenges posed by AI from two leading thinkers in the field.

4. **"Why The Future Doesn't Need Us" by Bill Joy (2000)**

- o **Publication:** *Wired*

- o **Description:** In this provocative article, Joy argues that developing technologies like AI, robotics, and nanotechnology could lead to a future where humanity is no longer necessary.

- o **Why Read:** It's a cautionary tale highlighting the potential dangers of unchecked technological advancement.

5. **"How to Stop Superhuman AI Before It Stops Us" by Stuart Russell (2019)**

- o **Publication:** *Scientific American*

- o **Description:** Russell discusses the challenges of controlling superintelligent AI and offers strategies for ensuring that AI systems remain aligned with human values.

- o **Why Read:** It provides practical insights into one of the most critical issues in AI safety.

Documentaries and Films

1. **"The Social Dilemma" (2020)**

- o **Director:** Jeff Orlowski

- o **Description:** This documentary explores the impact of social media and AI on human behavior, focusing on the ethical implications of data-driven algorithms that manipulate our choices and beliefs.

- o **Why Watch:** It's a compelling look at the real-world consequences of AI in our daily lives, particularly in the context of social media.

2. **"Do You Trust This Computer?" (2018)**

   o **Director:** Chris Paine

   o **Description:** This film examines the rise of AI and its potential risks to society, from job displacement to the creation of autonomous weapons.

   o **Why Watch:** It offers a balanced exploration of AI's promises and perils.

3. **"Ex Machina" (2015)**

   o **Director:** Alex Garland

   o **Description:** A thought-provoking science fiction film that explores the ethical dilemmas of AI, focusing on the relationship between humans and intelligent machines.

   o **Why Watch:** It raises important questions about consciousness, free will, and the potential consequences of creating indistinguishable AI from humans.

4. **"AlphaGo" (2017)**

   o **Director:** Greg Kohs

   o **Description:** This documentary chronicles the development of Google

DeepMind's AlphaGo, the AI that defeated the world champion of the ancient game of Go.

- ○ **Why Watch:** It's an inspiring look at the cutting edge of AI research and its implications for the future of intelligence.

Online Resources and Websites

1. **The Future of Life Institute (FLI)**

   - ○ **Website:** futureoflife.org

   - ○ **Description:** FLI works to mitigate existential risks facing humanity, particularly AI-related ones. The site offers articles, research, and AI safety and ethics initiatives.

   - ○ **Why Explore:** It's a valuable resource for staying informed about the latest developments in AI and efforts to ensure its safe and ethical use.

2. **AI Alignment Forum**

   - ○ **Website:** alignmentforum.org

   - ○ **Description:** A community-driven platform where researchers and thinkers discuss AI alignment, ethics, and strategies for ensuring AI systems align with human values.

- o **Why Explore:** It's an excellent resource for diving deeper into AI alignment's technical and philosophical aspects.

3. **OpenAI**

- o **Website:** openai.com

- o **Description:** An AI research lab focused on ensuring that artificial general intelligence (AGI) benefits all of humanity. The site provides research papers, blog posts, and updates on AI advancements.

- o **Why Explore:** OpenAI is at the forefront of AI research and development, making it an essential resource for anyone interested in the future of AI.

4. **Stanford AI Ethics Hub**

- o **Website:** aiethics.stanford.edu

- o **Description:** This hub offers resources, courses, and research on the ethical implications of AI, drawing from interdisciplinary perspectives.

- o **Why Explore:** It's a comprehensive resource for understanding the ethical challenges posed by AI and how they can be addressed.

5. **MIT Technology Review – AI Section**

- o **Website:** technologyreview.com/ai

- o **Description:** MIT Technology Review provides in-depth coverage of the latest AI developments, from technical breakthroughs to ethical debates.

- o **Why Explore:** It's a reliable source for staying updated on the cutting-edge developments in AI and its impact on society.

Conclusion: Continuing the Journey

Singularity represents one of the most profound shifts in human history, potentially redefining what it means to be human and how we interact with the world. The resources listed in this chapter are just the beginning of your exploration into this fascinating and complex topic. Engaging with these books, articles, documentaries, and online platforms can deepen your understanding, challenge your assumptions, and contribute to the ongoing conversation about the future of humanity and intelligence.

As we move closer to the Singularity, staying informed and engaged cannot be overstated. The decisions we make today shape the future, and by expanding your knowledge and perspective, you are taking an active role in shaping that future. The journey continues, and there is much more to discover.

Dave Karpinsky, PhD, MBA, PMP, is a globally recognized consultant, executive leader, and professional author whose work bridges business transformation, strategy, and personal development. With over three decades of experience advising Fortune 500 companies, government agencies, and high-growth startups where he traveled to more than 60 countries, Dave brings a rare blend of practical insight, operational excellence, and visionary thinking to every project—and every page.

His career spans top-tier consulting firms including McKinsey & Company, Accenture, SAP, Cognizant, BearingPoint, Ernst & Young, Infosys, and IBM. He has led multi-million-dollar strategic and technology initiatives for global leaders such as Capital One, Coca-Cola, Costco, DHS/TSA, Google, HP, Janus Henderson, John Deere, Lockheed Martin, McLaren, Merck, Nike, PetSmart, QuidelOrtho, and ViaSat, as well as large-scale public sector programs for the US Government, States of Alaska, Arizona, California, Florida, and Georgia.

As the author of numerous books on project turnaround, leadership, SAP implementation, and personal mastery, Dave is known for translating complex challenges into actionable strategies that deliver measurable impact. His writing combines analytical precision with compelling storytelling— whether he's decoding enterprise system failures or exploring the psychological dynamics of decision-making and influence.

Dave holds advanced degrees in business, technology and psychology, along with a portfolio of elite professional certifications. He is a sought-after speaker, strategist, and transformation advisor who empowers individuals and organizations to break through barriers and unlock lasting success.

Outside of his professional pursuits, Dave is an avid traveler and photographer, with a passion for astrophotography and a curated collection of high-performance and exotic cars. His global perspective, intellectual curiosity, and relentless drive to improve systems and people continue to inspire readers and clients alike.

To my constant joy and loyal hearts — you make life lighter

"We created artificial minds to serve us, but in teaching them to think, we forgot how to.

— *Dave Karpinsky*

www.ingramcontent.com/pod-product-compliance
Lightning Source LLC
Chambersburg PA
CBHW021539260326
41914CB00001B/73